HEROINES OF THE SKY

THE NINETY NINES, national club for women fliers, meeting at Valley Stream, Long Island, with Amy Johnson Mollison (standing at the extreme left), English flier, who had just flown the Atlantic with her husband. Amelia Earhart, first national president of the Ninety Nines, shown standing next to the English flier.

Heroines Of The Sky

By JEAN ADAMS & MARGARET KIMBALL

In Collaboration With JEANETTE EATON

Essay Index Reprint Series

BOOKS FOR LIBRARIES PRESS
FREEPORT, NEW YORK

TB 539
A3

STANDARD BOOK NUMBER:

8369-1539-9

LIBRARY OF CONGRESS CATALOG CARD NUMBER:

78-99615

PRINTED IN THE UNITED STATES OF AMERICA

☆

Acknowledgments

W E WISH TO ACKNOWLEDGE our gratitude to
the following for their kind co-operation: The National
Aeronautics Association of America, Washington, D.C.;
Hamilton Thornquist of the Boston *Evening Transcript;*
Leo Boyle, American Airlines, New York City; Nancy
Harkness Love, Inter-City Airlines, Boston; T. F. Hartley,
Boston Municipal Airport; Dan Rockford, *Time* magazine,
New York City; and Charles H. Gale, *The Sportsman Pilot,*
New York City.

Thanks are due to the following for permission to use the
material indicated:

Smith College *Alumnae Quarterly*—for "Caprice" by
Anne Morrow Lindbergh.

Anne Morrow Lindbergh—for "Height" which appeared
in *Scribner's Magazine,* April 1928.

Harcourt, Brace and Company, Inc.—for selections from
North to the Orient and *Listen! The Wind* by Anne Mor-
row Lindbergh; "Courage" from *Last Flight* by Amelia
Earhart.

v

☆

Foreword

Woman followed man into the air—yes, but she followed him so promptly that she may be said to be his partner in its adventurous exploration. When in the eighteenth century a Frenchman went aloft in a balloon for the first time, Madame did not tarry meekly on the ground. Only a year after this pioneer flight she caught up with Monsieur in the air. To be concrete, the name of that first heroine of the sky was Madame Thibaud and she was to thrill Paris audiences before the fall of the Bastille in 1789. Subsequently another Frenchwoman became so expert in her use of the balloon that Napoleon made her one of the chiefs of his air service.

In the case of the airplane, woman was tardier in joining her brother. Not until 1909—six years after Orville Wright made his historic flight—did the world greet its first woman pilot. She was the Baroness de la Roche, and immediately a number of other Frenchwomen entered the aerial lists. Soon women from other lands joined these first doughty pioneers. In the span of only one generation these modern

Valkyries have written upon the sky a record of courage, endurance, and imagination that forms the most romantic page in all feministic history.

The first woman passenger in an airplane was the French sculptress, Mme. Thérèse Peltier, who went up on July 8, 1908. Here it is inevitable that one should ask, "Why were they all French—these first ladies of the air?" Certainly it would seem more reasonable that the initiative should have been taken, not by France, where women have always been content to work behind the façade of Man, but by some land in which there existed an aggressive feministic movement. Perhaps the only answer can be found in a study of the race spirit of France. An affinity with the air is certainly part of that spirit, and it needs only a few facts to justify the statement.

Here is that handful of facts. The basic principle of the Zeppelin dirigible balloon was discovered by the Frenchman, Joseph Spiess. Another Frenchman, Louis Blériot, the first human being to fly across the English Channel, by that feat of 1909, inspired the first great wave of air enthusiasm ever felt in the world. What was more, Blériot refined the airplane, and his monoplane was the pet racer of the early part of the century. The earliest streamlined ships were of French origin. And when Claude Grahame-White, the famous English aviator, compiled in 1911 his list of the world's famous birdmen, France was represented by 387 names as against America's paltry 31.

Several months after Mme. Peltier made her ascent an

American woman was one of the crowd that watched Wilbur Wright demonstrate the airplane at Le Mans, France. She was Mrs. Hart O. Berg, wife of a Yankee businessman, and she was so thrilled by the performance that she asked Wright for a ride. Somewhat to her surprise, he consented, and a moment later she had jumped into the passenger seat. Seated there, she tied a rope about her ankles. It was the only way she could keep her long skirt from flying above her head, for one must remember that in those days there was no cockpit. The rider of the air sat on a perch no more sheltered nor commodious than a trapeze bar.

That first American woman who ever savored the airplane—incidentally, she was also the first woman to go up with Wilbur Wright—soared thirty feet and flew for two minutes and seven seconds. However, it is easy to imagine that she must have packed a lifetime of emotion into that short flight. For bumpy and wind-bitten was all travel in those planes described by one commentator as "85 per cent man and 15 per cent machine."

Outstanding among the women pilots of France in those early days was Mlle. Hélène Dutrieu. The daughter of an army officer, she was the founder of a long line of women who have never gone up without some pet talisman. Mlle. Dutrieu's amulet was a pair of gaiters worn by her father in the army. Protected by these gaiters, she made during the summer of 1910 one of her most spectacular flights. While the famous carillons of Bruges pealed out their welcome to her she circled above the cathedral of the old Belgian city.

Soon after this she came to America. The occasion of her visit was the first great international air meet ever held in the United States. In this contest, held at Belmont Park, New York, in October 1910, she was the only woman registered.

Just a month before the Belmont races an American woman had successfully tried her wings. On September 16, 1910, Dr. Bessica Faith Raiche, a Wisconsin-born physician who had been graduated from the Tufts Medical School in Boston, Massachusetts, took up a plane at Mineola, Long Island. That machine was an ugly biplane constructed by her husband, Mr. François Raiche. It was powered by a rebuilt two-cylinder boat motor which in her hands proved a fractious steed. She could get only a few feet from the ground on her first attempt and later trials ended in bad landings. Nothing daunted, however, she tried again ten days later. This time, according to a contemporary record, she made "an undeniable flight, driving with great coolness and judgment." By this success she won her title. Dr. Raiche, who died in 1932, was the first American woman ever to pilot an airplane.

Almost simultaneously, however, another American woman won her spurs. On October 22, 1910, Miss Blanche Stuart Scott took up a plane at Fort Wayne, Indiana. Miss Scott, the only woman pupil of Glenn Curtiss, pre-eminent American rival of the Wright brothers in the early lists of airplane construction, always had a cordial attitude toward adventure. After a career in aviation, spiced by a fall in which she broke twenty-three bones, she earned another title. She was the first woman to drive an automobile across the continent. That tour, during which the governor of each state she visited gave her an official welcome, was widely publicized. Yet the automobile never could replace the airplane in her heart, and during later years we find her using her pen for the promotion of air-mindedness. Various articles on aeronautics and the development of commercial aviation led to her appointment in 1929 as head of an educational division of the Maximum Safety Airplane Company.

However, grateful as we are to these two pioneers, we must acknowledge that not until 1911 did woman's career in the air achieve a truly professional character. Neither Dr. Raiche nor Miss Scott had received a license. And it was only when Harriet Quimby got such recognition from the Aero Club, which at that time assigned all pilots' licenses, that the line royal of great women flyers was really begun.

Miss Bernetta Miller of Canton, Ohio, who was the fifth licensed woman pilot in the United States, secured a unique place in our feminist record. On that summer day of 1912, when she was scheduled to make her test flight, the wind was so·strong that she could not go aloft. But the moonlit night of that same day was calm and so Miss Miller became the first woman to qualify for her license at night. Later on she was to win another "first." When the Government wanted a demonstration of the monoplane, she was chosen for the job and, although she failed to win official Washington from the biplane, the fact that even before World War I a woman was entrusted with such responsibility is significant of the strides taken by the aviatrix in a few short years.

Among the more piquant characters of that second decade of our century was Mrs. Hilder Florentina Smith, an Illinois woman first heard of as a member of a flying aerial team called The Flying Sylvesters. In fact, Mrs. Smith was as addicted to groups as the family album itself, for in July 1912 she was a passenger on the first family cross-country flight in this country. That flight from Santa Ana to Los Angeles was followed by the exhibition tour which she and her husband made through the Middle West.

Of course Mrs. Smith was capable of brilliant solo work, as is proved by her hop in a Martin pusher biplane in 1914. Even more spectacular was her part in the opening of the Los Angeles Harbor in April 1914, for at that time she made a parachute jump from a plane piloted by Glenn Martin.

Mrs. Smith made a contribution to aviation even more arresting than her flying exploits. She became interested in the mechanics of the plane and was to help her husband, Floyd Smith, construct the first Floyd Smith tractor biplane in 1912. For this reason, her name must be linked with that of another woman. In those pioneer days people who knew Miss Lillian Todd found her always bending over a drawing board. She was designing an airplane. So great was the faith of her friends in her design that the late Mrs. Russell Sage was persuaded to finance this, the first airplane model ever essayed by one of her sex.

Alas for such hopes! No satisfactory engine could be found for the body and afterward we hear of Miss Todd launching a novel enterprise. It was a school for boys who wanted to learn how to build airplane models. This novel form of education had its home in Los Angeles where Miss Todd, together with Blanche Stuart Scott and Ruth Law Oliver, constituted a little colony of pioneer fliers. It is interesting to note that Miss Todd's scrapbook, culled from aviation news since Wilbur Wright made his first flight at Kitty Hawk, is now in the Los Angeles Museum.

Natural enough that death should have taken its toll from the pioneer women who braved those airplanes which were "85 per cent man and 15 per cent machine." The fall of Harriet Quimby into the waters of Boston Harbor in 1912, that tragedy which put an end to the brilliant flying career which had begun only a year before, dominates the somber record. Yet she was not the first American woman to give her life because of love of the air. Miss Denise Moore was that. In 1911, the very year Miss Quimby was studying at the Moisant Aviation School in Long Island, Miss Moore was a pupil at the Farnum School in France. Ready for her license test in three weeks' time, she crashed and died while taking that test. The next year, 1912, which witnessed the death of Harriet Quimby, mourned also Mrs. Julie Clark, our third licensed pilot. Seven days after her qualifying flight she came down at Springfield, Illinois.

It was not only in actual battle with the air that the woman of yesterday made her contribution. Orville Wright is quoted as saying, "When the world speaks of the Wrights, it must include our sister. Much of our effort has been inspired by her." This statement is amplified by Amelia Earhart who wrote that Katherine Wright not only acquired Greek and Latin but gave the money she earned as a teacher in these subjects to her brothers so that they might continue the aeronautical experiment which by this time occupied them to the exclusion of bread-and-butter business. And thus helped pay for, and helped build, the first heavier-than-air plane ever flown.

In spite of this reminder, few people in the world remember that the Wright brothers had a sister. An oblivion almost as profound has been the lot of our early women pilots. How many of us today know the name of our first licensed aviatrix or realize that she was the first of her sex to fly the English Channel? Which of the younger generation can place Katherine Stinson and Ruth Law as being queens of the air during the World War period of 1914? Indeed, many fliers of much more recent years have become almost as shadowy as those dauntless pioneers who rode the air, not in today's comfortable cockpit, but between the struts of their skeleton planes.

It must be admitted that the men who have written about aviation have not been helpful in preserving the memory of America's heroines of the air. Typical of masculine reticence is the *Aircraft Year Book* of 1919. Although women had contributed much to aviation during the eight years which preceded the publication of this record, not a single woman flier is mentioned. With a brief notation on the women employed in airplane factories, the subject is dismissed.

Because of such negligence on the part of men authors and men recorders, we have been obliged to consult many scattered sources in our attempt to evaluate the part played by American women in the great drama of the air. Newspaper "morgues," old magazines, occasional bits in some biography—these have supplied our hints as to many of the fliers active during the fifteen or sixteen years which followed August 2, 1911—the date which greeted our first woman pilot. It goes without saying that these sources were often at complete variance and that a tremendous amount of sifting has had to be done in order to approximate the truth. In this work of sifting we wish to acknowledge our debt to America's greatest and most representative club of women fliers, the Ninety Nines.

Aside from such sources and always dominating them is one authentic record. Realizing that the skill of pilots must always be rated by an unbiased judgment in which the whole world has faith, the French, early in the history of aviation, founded an organization dedicated to the establishment of standards. This organization, called the Fédération Aéronautique Internationale, measures, not only the performance of fliers, but of various planes. And, in order that the international rulings may be executed, the organization has a representative body in every country.

In America the first representative of the French society
was called the Aero Club. Later this club merged with another to form the National Aeronautical Association. Today the N.A.A. always has the final word. Not a single race
can be run, not a single contest can be held, not a single
attempt at setting a new record is recognized, unless this
organization first gives its sanction. Furthermore, its representatives must always serve as judges and issue the final
official ratings.

Affiliated with the French Federation is the International
League of Aviators. And what is the function of this
branch? It is twofold. It stimulates interest in aviation and
also initiates various sporting events. A contest committee is
the center of its activity, and about a decade ago it made
room for one woman member. This woman is today always
chosen by the Ninety Nines in recognition of the outstanding merits of one of its members. The International League
awards each year the most coveted of trophies, the Clifford
Burke Harmon Trophy which is given to the most eminent
sportsman flier—either man or woman—of the current year.

Manifold are the types of American women who have
heeded that one imperious cry, "Go aloft." Professional
writers and actresses, trained nurses and dancers, business-
women and college girls, daughters who would ordinarily
remain in the unadventurous setting of their own families,
and mothers who at one time never dreamed of the air
as a vocation—all are represented in our list of famous
aviators. They have come from all parts of the United States.

They can boast such personal beauty as belongs to Louise Thaden and such versatility as belonged to the late Amelia Earhart. They have drawn to one common level, the air, that scholarly master of poetic prose, Anne Lindbergh, and that practical sculptor of her own fate, Jacqueline Cochran.

To dismiss such widely divergent ' temperaments and backgrounds with mere statistics about flights, altitudes, and endurance is obviously unsatisfying, so we have attempted to present history in the form of a portrait gallery. If we have been able to give something of the personality which lies behind each flier's record we shall feel that our goal has been reached.

A portrait gallery! Yes, but through every canvas there runs one unmistakable resemblance. Supreme courage— that is the common stamp. To all of them may be applied in some measure the tribute which Ruth Nichols once paid Amelia Earhart's flight across the Atlantic. "Here," said this distinguished flier, "showed, not the old hereditary courage of women, the instinct to meet unflinchingly some sharp crisis of pain. Here was the enduring courage of man."

Many of these modern Valkyries have not dedicated their bravery merely to making some new record. The gallant work done by both Katherine Stinson and Ruth Bancroft Law during World War I; the broad humanitarian scope of Relief Wings, that organization so ably directed by Ruth Nichols; the archaeological research of Anne Lindbergh in Central America; the invaluable aid Phoebe Omlie is now rendering her government; Jacqueline Cochran's organization of twenty-five outstanding American women pilots to ferry planes from English factories to English training centers—all these achievements indicate something much deeper than spendthrift and vainglorious courage. And yet they are merely blazing more glorious paths of service for a younger generation of women who will one day ride the clouds.

THE AUTHORS.

March 31, 1942.

Contents

List of Photographs

Harriet Quimby and Mathilde Moisant

1911 GREETS OUR FIRST GREAT ONES

☆

THE SKY above Belmont Park on that Sunday afternoon of October 1910 was sunny, but its few clouds raced before a stiff southerly breeze. The wind smacked the flags—the Stars and Stripes, the tricolor of France, Great Britain's Cross of St. George which, together with other national emblems, dotted New York's famous racetrack. It rippled the fronds of ostrich plumes worn by women spectators in that crowd of fifty thousand people. What was more, it furrowed many a brow. How could any aviator possibly fly to the Statue of Liberty and back again in a wind like this? Yes, it was doubtful now whether they could pull off the great race which offered ten thousand dollars to the winner.

This day of October 27 marked the end of America's first great international air meet. For a week it had brought together famous birdmen from all over the world. For a week throngs recruited from every level of society had thrilled to the exploits of these men—Claude Grahame-

White, the Englishman idolized by all American debutantes; Count de Lesseps and Leblanc, two of France's able fliers; John Moisant, the Chicagoan of Latin descent; and J. Armstrong Drexel, handsome young English-bred scion of Philadelphia's banking family.

Only yesterday crowds had cheered when Claude Grahame-White had carried off the speed race with something above sixty-two miles an hour. Ah yes, but if it had not been for a mischievous telephone pole—— How about the Frenchman Leblanc? For he had actually been flying at sixty-eight miles an hour when the accident disqualified him. That was really the big news of the week, and it filled the air with prophecy. Sixty-eight miles an hour! Why, who could tell—at this rate the airplane might some day get up to eighty miles an hour.

But, thrilled as were the crowds by the strides taken in the new sport, our native pride was touched. What if the airplane had been born in America seven years before? Now foreign aviators were carrying off all the honors. At the beginning of the tournament Belmont had pinned its faith upon John Moisant. However, that daredevil jockey of the air, first to carry a passenger over the English Channel, had met with various mishaps. At the very outset his plane had been wrecked by a high wind. And yesterday in the speed race—— True, he had won second honors, but distinctly his machine was inferior to the Blériot racer used by the victor, Claude Grahame-White. As for today, even if the others decided to dare the wind—— No, John Moisant

would not be among those who winged to the Statue of
Liberty. Where would he get a plane that would enable him
to compete with the English and French?

The crowd in the grandstand fidgeted. Every eye was
fixed on the sheds sheltering the various planes. Was anybody going to risk it? It was something past three when
there was a sudden flurry over there where the tricolor
rippled in the breeze. Mechanics began to pull from its rude
hangar a monoplane.

As they did so a girl in the grandstand who had been
watching through field glasses suddenly caught her companion's arm. "He's going up—Count de Lesseps is going
up!" she cried excitedly. "Come on, let's walk over there
and watch! I can't half see from here!"

"Oh, Harriet," pouted her companion, "can't you ever get
close enough to those old machines? I believe you've fallen
in love—yes, and his name's Airplane."

Suddenly Harriet Quimby's lips parted. It was only then
that you realized fully the great beauty of this young Cali-
fornian who was dramatic critic for that mellow periodical,
Leslie's Weekly. For her smile not only revealed even white
teeth, but it caught up her whole face into a look of such
radiance that people thought of some young goddess here
on earth for a mere sojourn among mortals. "Maybe you're
right," she assented gaily.

A moment later the two young women were walking toward the spot where Count de Lesseps was preparing to
take off. The long skirts of their tailor-made suits were

tugged by the wind. The jabots of their lingerie blouses blew out like banners. They both held their big-crowned hats to their heads.

"Heavens!" cried Harriet's companion as they came to a standstill, "anybody's a fool to go up a day like this. I should think it would be a lesson to them—what happened to John Moisant in that wind the other day."

At mention of Moisant's name Harriet's face clouded. "What a shame he can't get into the race today! Oh, if he only had a decent plane I'm sure America would have a chance!"

Her companion gave her a teasing glance. Then she began to hum "My Hero" from the current musical hit, *The Chocolate Soldier*.

"Well, he is romantic—you'll admit that." Thus Harriet defended her hero worship. "Think of his being in those revolutions down in South America—always on the losing side; getting himself in jail—for all the world like a Richard Harding Davis hero. But look"—her eyes had fallen on a short, stocky girl wearing a combination of overall and jodhpur—"Mademoiselle Dutrieu! I've been wondering where she was today."

With distinct disapproval her companion eyed the only woman flier at the air meet. "Yes, and what a sight she looks. If I had to wear things like that every time I left earth, I'd let the towers of Bruges live alone for the rest of time."

"Oh, how silly of you!" Harriet exclaimed hotly. "I never look at her without thinking what a superb thing it was to do—to fly around the Bruges Cathedral. Think how she must have felt up there with all the bells pealing around her—just as if she were the 'lark at heaven's gate.' Oh, why is it that American women are so slow? Haven't we any nerve over here that we allow the French to beat us?"

"Hmph! For a magazine writer you certainly don't keep up very well with the news. Didn't you read about the woman doctor who piloted her own plane at Mineola last month?"

"Oh, Doctor Raiche! Yes, of course I read about her. But —well, that's just kind of amateur sport. Compare that with the Frenchwomen having their licenses and everything— going to all the big air meets—doing the same things that the men do."

By this time the mechanics who had been testing the plane gave it their final approval. A moment afterward Count de Lesseps climbed into his seat, that open perch which provided a board for the feet. As he did so, a young fellow in the crowd called out, "Some close fit, isn't it? Good thing it isn't President Taft who's trying to sit there."

There was a roar of laughter in which even the French aviators joined. They had been in this country long enough to absorb the fact that nothing could possibly be funnier than a joke about President Taft's weight. A minute after came the whir of the plane's propeller. Then, slowly, Count de Lesseps mounted. He was the first to set out for the Statue of Liberty, that somewhat distant lady who waited for him more than seventeen miles away. And as he rose cheers burst from fifty thousand throats.

With entranced eyes Harriet followed him—first the wind-blown man's figure sitting there between the struts, then a mere dark blot, then a nothing absorbed in the triumphant, impersonal glint of wings. As she watched, her lips parted. The eagle soaring to its mountain crag—the spirit of everything which could loosen Earth's bonds. Perhaps it was at this moment that Harriet Quimby first knew what she most wanted in this life.

It was not long after this that Claude Grahame-White, the English aviator, followed Count de Lesseps. Again a pang went through Harriet's heart. The French and the English—yes, they were both competing for the ten-thousand-dollar prize. But America? Disconsolately she and her friend went back to the grandstand.

A boy in front of her humming stubbornly "Every Little Movement Has a Meaning All Its Own." A woman behind her saying how wonderful it was that Sarah Bernhardt was back in America—"And, my dear, they tell me you never notice it—that leg, you know." Somewhere a voice exclaiming, "There she is—Mrs. Drexel. Yes, stunning, but she'll never hold a candle to her mother. I assure you, Mrs. George Gould was one of the most exquisite creatures in the world." Somewhere another voice talking about Hawley and Post, the long-lost balloonists who had so miraculously shown up. Then Harriet turned around. So did everyone else.

It was the far-off buzz of an engine which had galvanized

everybody. Immediately the buzz became a speck. The speck was advancing steadily. What! A Blériot monoplane. But whose could it be? Tensely the crowd watched and waited. Then a mighty cheer recognized the man at the wheel. "Moisant," bellowed the men, and the women waved their handkerchiefs. In some way the daredevil had managed to get a plane. He was going to enter the race, after all.

Through her glasses Harriet caught the small figure which descended from the plane. He was grinning, and his teeth flashed vividly against a coppery red skin. Ah, no wonder London had gone mad over him this past summer —this jockey of the air with his handsome face! She was sorry when a throng of people blotted him from view. And a moment later she had dragged her reluctant friend to the field.

They stood while he prepared to take off. When he finally rose from the ground Harriet's eyes followed him even more rapturously than they had watched De Lesseps, for she had learned by this time why John Moisant's flight was so much more daring than that of his competitors. He had just bought that Blériot monoplane in Brooklyn. He had never flown it before, and here he was, on this windy day, taking it out over the harbor.

By this time both De Lesseps and Grahame-White were back. The Frenchman had made the better time. Not a doubt about it. He was the winner. That there was any real chance for Moisant—nobody dared hope this. But when strong field glasses brought the first sight of his returning

plane, a ripple passed through the throng. It grew stronger
as he neared. Tensely people looked at their watches. At
this rate—— But could it be? In the plane he had never
flown?

Thirty-four miles in about the same number of minutes.
It was true. Watches couldn't lie. John Moisant had won the
race. America had won the race. In their frenzy men rushed
at the small young man and wrapped him in the American
flag. Then they carried him around the field on their shoulders. Before this, however, two girls, both dark and slight
as he himself, had rushed toward him with outstretched
arms. Quite evidently they were his sisters, and Harriet regarded them with a little envy. Think of being related to
John Moisant!

The words which we have put into the mouth of Harriet
Quimby are not historical. Nevertheless, they are justifiable,
for they interpret facts which are a matter of history. She
did go to the Belmont races of 1910; she did fall in love with
the airplane at first sight; and she did make up her mind
then and there to learn to fly.

It was not until the following summer that the young
magazine writer—she was now twenty-seven—really set
about the fulfillment of her ambition. In the meanwhile,
John Moisant had crashed to his death at New Orleans and,
though she had never met him in person, she shared the
grief of the thousands who had thrilled to his exploits. One
of the legacies which Moisant had left was a flying school at

Garden City, Long Island, and it was here that she applied for instruction.

"I suppose it seems a little odd—for me—a woman," she began falteringly to the young man who received her.

"Not at all." He smiled back. "The fact is that Miss Mathilde Moisant has already started lessons."

Mathilde Moisant was one of the two slight, dark-haired girls who had rushed forward to greet their hero brother after he returned from his successful flight to the Statue of Liberty. Harriet's mind flashed back to the Belmont races. Perhaps she may even have felt a little stab. Mathilde Moisant had already started lessons. This meant that she herself would not be the first qualified woman pilot in the United States.

Might *Leslie's Weekly* object to her new enthusiasm? Might they fear that flying lessons would distract her from her work? Fearful of this, the girl arranged to have her lessons at dawn. Her instructor, a young Frenchman named André Houpert, was also teaching Mathilde Moisant, and very soon he arranged a meeting between his two women pupils. Under any circumstances the two young women would have liked each other, but friendship was cemented quickly by their one common enthusiasm. Propellers and engines, struts and wires—such terms were shuttled between them as fervently as are proms and football games between two ordinary girls. Yet underneath was the rivalry to be expected from friends who have fixed their affections on the same object. Which would be the first to receive her license from the Aero Club?

For some unexplained reason it was Harriet who won. August 2, 1911—this date should be memorized by all who care about women's achievements in the air. Then for the first time America could boast a licensed woman pilot. The honor had come to Harriet Quimby after thirty-three lessons involving four and a half hours in the air.

It was two weeks later that Mathilde received her license. To be exact, it was on August 17, 1911. To the pretty, bright-eyed brunette this date was infinitely less precious than her trial day, August 13, for was not thirteen her lucky number? Born September 13, 1886, she straightway proceeded to name her plane Lucky Thirteen.

On her twenty-fifth birthday, September 13, 1911, she took out Lucky Thirteen for a cross-country flight. In those days every flier made this gesture. It was just like saying to the world, "See, I'm a real aviator." In Mathilde's case the gesture was convincing, for during this flight she set the women's altitude record. It was 1,500 feet! Superior modern lips may smile at this height, now hardly the minimum required for passage over towns and cities, but thirty years ago—what a flurry among newspaper typewriters!

"Sister of the late John Moisant takes long glide through the clouds, then volplanes down, makes her way up again to a high mark, and at last shuts off her motor for a toboggan down the air lanes"—thus a contemporary reporter described the event.

After this exhibition Mathilde Moisant's name became known all over the country. Then suddenly she extended her fame. She became the first woman ever to fly over Mexico City. The date? Why, of course, it was on November 13 of the same year. For this feat she received $490 in cash.

Meanwhile, we may be sure of one thing. The ambitious Harriet Quimby was not idle while her friend was making a name for herself. Just two days after she received her license we hear of her making a moonlight flight at Dongan Hills, Staten Island. Twenty thousand people were there to gasp over the exploit, for was not this the first time any American birdwoman had turned nightingale?

Still greater crowds turned out to watch her subsequent exhibition flights at the Richmond County Agricultural Fair. It was then that she first donned the spectacular flying costume in which she later made all her important flights. The trousers tucking into high laced boots as well as a blouse with long sleeves, choker collar, and a monklike hood were made of mauve-colored satin! By no means was this garb typical of the day. Did not her friend Mathilde Moisant fly in a divided skirt and Mademoiselle Dutrieu elect trousers which were a cross between jodhpurs and overalls? Undoubtedly some of her rivals sniffed, "Does she think she's going up in the air or going on the stage? That's what comes of being a dramatic critic."

Did she ever dream, when she stared enviously at Mlle. Dutrieu at the Belmont races, that in less than a year's time she, a comparative novice, would be competing with the stocky French airwoman? It must, indeed, have seemed stranger than fiction—that day at Belmont Park, Long Island, when Mlle. Dutrieu appeared as challenger of the women's endurance record which the American girl had just made. However, Harriet was not to defend her title. Why? Because the day on which she was to meet the challenger happened to be Sunday. In vain she was pressed to go up by various friends. Her one answer was, "I can't. I promised my father I'd never fly on Sunday." Later on she was to see the French challenger beating her own record. That victory brought to Harriet's face only a quiet smile. "I don't care," she said. "To me a victory won on Sunday is not worth while."

Her successful French rival must have been more than a little amused by this puritanical attitude. Sunday? *La, la, la,* what was it for except opera and theater, cafés and parties? Mathilde Moisant shared this Latin feeling about Sunday, and once she defied the "blue laws" in a way that made history. The scene of that defiance was at Mineola, Long Island, where she and some men aviators had been engaged to appear on a certain Sunday. Immediately, however, they were warned by a belligerent sheriff that anyone who went up would be arrested. Undecided, the performers hung about their hangars. Then one daring male took off. Soon another followed him. And at this point Mathilde showed herself the true sister of John Moisant, insurgent fighter in South America.

"So I'll be arrested, will I?" she cried in a voice shrill with excitement. Then she turned to her mechanic. "Tune her up, Mike," she commanded.

The big crowd which had gathered on the field cheered lustily. It cheered even louder when she mounted. And in the ensuing minutes it had a thoroughly good time. For was there ever a game of tag so grotesque as this one? Mockingly the elfin little figure in the plane darted from one end of the field to the other. Panting and swearing, the sheriff and his deputies tried to follow on foot. Oh, wait till she came down, the impudent little black-eyed chit!

However, she didn't choose to come down. She winged toward her brother's field near by. There the officers of the law were waiting for her. So was a big throng of Moisant sympathizers. When the deputy sheriffs tried to seize her, a riot ensued. Clothing was torn and faces scratched. But at last the valiant Mathilde escaped. And that evening she was completely vindicated.

"She was violating no law," said the justice of Hempstead when the deputy sheriffs appeared before him to ask for the arrest of Miss Moisant. "She had as much right to fly an airplane in the skies on Sunday as anyone has to use the ordinary highways with automobile or carriage."

Again the fighting Moisants had scored a point. After this there was little agitation about Sunday flying in the New York fields.

By the autumn of that year Mathilde and Harriet were

making joint exhibition flights over Mexico and various parts of America. Thousands always assembled to watch them—the gorgeous bird of paradise in her mauve-colored satin and the brave little sparrow in her divided skirt. Sometimes, too, the audience got an unexpected thrill. Once Harriet's plane balked when she was only a hundred and fifty feet above the ground, and it was skillful maneuvering alone which enabled her to clear the obstacles below for a successful landing. The very next day something even more serious happened to Mathilde: Her plane smashed, and with difficulty she escaped from its wreckage.

"Good old Lucky Thirteen," she may have murmured as she pulled herself out.

And with equal likelihood we hear Harriet's answer: "Better not depend too much on your old thirteen. You're awfully reckless—you know you are, Mathilde."

However, even Harriet, much more careful than her friend about going over her machine before starting, was not completely above superstition. She always wore with her mauve-colored satin an antique necklace and bracelet which were supposed to bring good luck. Then came a day when she added to these amulets.

"See," said a French aviator to her as he drew from his pocket a little brass god, "I'm going to throw him away —this *petit monstre*. He brings me no luck. He is *méchant* —what you say—bad?"

"Oh no," protested the girl. "Don't throw him away. I'll

take him—your little god—and make him behave. You'll see, I shall carry him and he will bring me good luck."

By this time, of course, *Leslie's Weekly* was fully aware of the spectacular role which their dramatic critic was playing in the air. Far from discouraging her, they played up her activities, and when she undertook the most ambitious exploit of her career they made only one stipulation. She should give them exclusive American rights to a first-person story of her adventure.

And what was the adventure to which Harriet had now set herself? It was the crossing of the English Channel. Ever since Blériot spanned it for the first time in 1909 almost every famous man flier had followed him. Then at last the aviator's supreme goal—a nonstop flight between London and Paris, that 290-mile course which so many airmen had tried in vain, was gained in 1911. So far, however, no woman—not even the doughty Mademoiselle Dutrieu—had crossed the Channel. To Harriet Quimby, ever ambitious to make her sex man's peer in the air, an offer from a London periodical to finance the exploit proved irresistible. By this time, of course, she had a manager, and it was with him that she sailed for England in the spring of 1912.

A word right here about the strides taken by aviation since the October days of the Belmont air races. Then the world had gasped when the Frenchman Leblanc flew at 68 miles an hour. Very soon after this, however, a compatriot of his had bettered the record by reaching 82.7 miles.

The name of this second Frenchman, Nieuport, should always be venerated in aviation, because the machine in which he made this record was the direct ancestor of modern streamlining. He had used a flatter wing curve and refined the shape of the fuselage cover. And he was merely paving the way for another French model which appeared in 1912—a model which attained the staggering speed of 108 miles an hour.

As her steamer plowed through the gray oily stretches of the Atlantic we can feel across the span of thirty years the pulse of this American girl sailing for England in 1912. She was going to fly for the first time in her life that racer preferred by nearly all famous men fliers—a Blériot monoplane. She was going to fly—also for the first time—across water. She was going to dare what no woman had ever dared before. Ah yes, there was reason for the dazzling smile which she so often turned upon her fellow-travelers, those poor, underprivileged people who were going abroad merely for sight-seeing and shopping.

Naturally, she was all eagerness to try out her new plane. England, however, was not the place to do it, for absolute secrecy about her plans must be preserved until the moment she set off. What if some other woman, getting wind of the Channel crossing, should beat her to it? Consequently, she took that shining new Blériot to a remote and unpretentious French resort called Hardelot. Here she would get her practice before the great flight.

Alas for any such hopes! A high wind haunted Hardelot

to shake the flimsy little villas and lash the sea into mountainous waves. Pacing up and down her room, gazing from the tormented sea to the wind-filled sky, she waited for several days. Impossible to go up in weather like this. Yet time was pressing. Yes, there was only one thing to do. She must go back to England and make her crossing in a plane she had never tried.

Never before this time had she used a compass. But now, since everyone insisted that it would be sheer madness to attempt the Channel without this instrument, she received some instructions in its use from an aviator named Hamel. It was Hamel, in fact, who sent her off on her adventure with the solemn admonition, "Whatever else you do, Miss Quimby, be sure to keep to your course, for if you get five miles out of the way, you will be over the North Sea and you know what that means."

Well indeed she did know it. But if any fearful thought of the men who had been blown out to those deathful waters was in her mind she did not show it on that morning of April 16, 1912, when she stood on the white cliffs of Dover, for her characteristic smile of the young goddess was on her lips as she surveyed the group which had gathered about her. Over there lay the coast of France—she would think of nothing else.

She and her mechanic went carefully over every part of the plane. Then she got into her open seat—the cockpit was not to come for several more years—and waved her hand. There was a last flashing smile and then she was off—the dazzling violet apparition—without parachute. Without any of the modern instruments. In a plane which was hardly more than a winged skeleton with a motor on it; one, furthermore, with which she was totally unfamiliar. As Amelia Earhart suggests, we must consider all these things in measuring the achievements of Harriet Quimby. To cross the Channel in 1912 required even more bravery and skill than to cross the Atlantic today. Always we must remember that in thinking of America's first great woman flier.

And how did she fare in her hazardous trip? Let Harriet Quimby tell you herself. The following is a quotation from her own description of her trip as it appeared in *Leslie's Weekly:*

I was hardly out of sight of the cheering crowd, before I hit a fog bank and found my needle of invaluable assistance. I could not see above, below, or ahead. I ascended to a height of 6,000 feet, hoping to escape the mist that enveloped me. It was bitter cold—the kind of cold that chills to the bones. I recalled somewhat nervously Hamel's remark about the North Sea, but a glance at my compass reassured me that I was within my course. Failing to strike clear air, I determined to descend again. It was then that I came near a mishap. The machine tilted to a steep angle, causing the gasoline to flood, and my engine began to miss fire. I figured on pancaking down so as to strike the water with the plane in a floating position. But, greatly to my relief, the gasoline quickly burned out and my engine resumed an even purr. A glance at the watch on my wrist reminded me that I should be near the French coast. Soon a gleaming strip of white sand flashed by, green grass caught my eyes, and I knew I was within my goal.

Yes, she was within sight of her goal, and a few minutes later she was to touch it. She came down near Hardelot, the same wind-blown French town where she had vainly hoped to test her Blériot. There she was welcomed by a throng of wondering fisherfolk, who warmed her with tea and tried desperately to make her understand their torrents of French. It was not until she got back to England that she was to hear that another aviator who had attempted to cross the Channel that day had been lost. He had been pulled out of his course and over the North Sea.

A heroine! Yes, she became that both in England and in America. But undoubtedly some of the glory of her feat was obscured by a tragedy which dwarfed all other news. Shortly after she crossed the Channel the luxury steamer *Titanic* struck an iceberg and was sent to the bottom. The great ship took with it many distinguished people to an ocean grave, and it was natural that for weeks people could talk of nothing else.

Once back in America, she slipped into her routine duties at *Leslie's Weekly*. But doubtless she felt now a sense of anticlimax. She had performed the most perilous aerial feat ever undertaken by a woman. Now what? Shuttling between her office and the Victoria Hotel where she lived with the devoted mother who was so proud of her success, she sighed for new worlds to conquer. From these restive days there springs a legend. One day when everything seemed to go wrong she took out the little brass god which the French aviator had given her and which had accompanied her on her trip across the Channel. "Horrible little monster!" she cried, and cut off its head.

If this legend be true, she defied Fate. Perhaps she defied it even more by an article which she wrote for *Good Housekeeping*. Turning back to the 1912 files of this magazine, you will find an article signed by Harriet Quimby. The airplane was an ideal sport for women and—oh, so safe. "Only a cautious person—man or woman—should fly," she wrote. "I never mount my machine until every wire and screw has been tested. I have never had an accident in the air. It may be luck, but I attribute it to the care of a good mechanic."

A good mechanic was with her on that day, July 1, 1912, when she entered the air meet at Boston. He had gone over every part of her machine. Serenely she watched him. Even so, she may have touched her constant amulet, the antique necklace which, as usual, she wore with her costume of mauve-colored satin. Yes, and she may have remembered the decapitated brass god under her blouse. All right now. Ready. She turned her old dazzling smile to the thousands of people. Then she mounted the pilot's seat. Even far away you could catch the violet-colored silk of her costume. Vivid as a Venetian sail it stood out against the flaming gold of the sunset sky.

She carried with her a passenger. He was Mr. William Willard, manager of the air meet. The audience, however, hardly noticed him until—— A beautiful ascent above Boston Harbor followed by a swift downward swoop, then—— Horror glazed every eye that watched. The passenger had catapulted from the plane. Another second and, outlined against those sunset clouds, Harriet fell to the water beneath. She was gone—the beautiful young Phaëthon—gone eleven months after she had received her license.

Nobody could ever explain the cause of the tragedy. Had something gone wrong with the engine? Had she fainted as she dived so swiftly from her five-thousand-feet altitude? Another guess seems more plausible. This is that contact with a pocket of air dislodged her passenger. Since he replaced the bag of sand which her Blériot monoplane required for balance, his fall may have so upset this balance that she, too, was jerked from her seat.

Whatever the cause of the tragedy, America mourned her deeply—the first of our great women aviators. The New York City newspapers which had failed to give her front-page space at the time of her Channel crossing made up for the omission now by lengthy accounts of her career. *Leslie's Weekly* paid her eloquent tributes in their obituary. In this chorus of lamentation we catch across the years one particular note which brings both a smile and a tear. Her devoted mother could not forget Harriet's satin suit and her antique jewelry. Some vandal had stolen them while her daughter's body lay in the morgue. "And I wanted the Smithsonian Institution to have them," sobbed the grief-stricken old lady.

And Mathilde Moisant? She was flying in formation with her friend when the tragedy occurred. She saw the two bodies whirled to the water, and it took all her will power not to faint. Yet, true child of the fighting Moisants that she was, she managed to guide her plane to earth.

Several months afterward she was to express to a reporter what she felt about her loss. "How I wish somebody could tell me it wasn't true—Miss Quimby's death," she said. "No accident except that to my brother Jack ever affected me so much." And then she added, half sad little smile, half shrug, "When I think how she was always scolding me for my carelessness, and here I am after all my accidents while she had to die in her first mishap. Well, after all, flying's like that—just a game of poker. You're always confident that, no matter what happens now, next time you'll win."

Certainly if there were ever an authority on accidents it was this same little Miss Moisant. We have spoken of the wrecked plane from which she escaped while she and Harriet Quimby were giving exhibition flights in Mexico. This was merely a forerunner of other narrow escapes. It was some weeks after Harriet Quimby's fatal fall that she had another major accident. Giving an exhibition at Shreveport, Louisiana, she struck a hummock. The machine turned a complete somersault and she was pinioned beneath it.

HARRIET QUIMBY, first licensed woman pilot in America, and first woman to fly the English Channel, in her famous violet satin flying costume.

MISS QUIMBY's plane, typical of the planes in use in 1912.

Brown Brothers

Brown Brothers

MATHILDE MOISANT, second woman to receive a pilot's license in this country.

Brown Brothers

"I escaped with a bruised face," she reported merrily. "I think it was my size that saved me. But oh, if it had been a biplane, then nothing could have saved me."

A few days later she was again flying at Wichita Falls, Texas. No memory of previous mishaps clouded her face. True enough, flying was a game of poker, a game which obliterated the past and brought you only hope of the present. A big crowd had assembled to watch her ascend, and, when she attempted to land, that eager crowd pressed forward. In a second she was given her choice. Should she kill people in that crowd or should she take a chance on her own life? For brave Mathilde Moisant the decision was easy.

She turned on the spark and tried to ascend. But the machine shot up only thirty feet before it came down. It was with such force that the oil tanks were loosened. Immediately the whole ship was ablaze. The big crowd rushed forward. They did not expect to see Mathilde Moisant alive. They stared at a tiny figure which emerged from the blaze as if they were beholding a ghost. Only her hair and leggings were singed. Was this the work of her faithful thirteen again?

"Will you please telephone my sister?" said Mathilde Moisant quietly. "She may be worried about me."

That was her final appearance in the air. It was not, however, because she lost her courage that she retired. These Moisants, like all Latins, had an intensely tribal feeling. They loved each other even more than they did danger, and it was because Mathilde could no longer endure the suffering she imposed upon her family that she did as they wished. She retired to a ranch. At the age of twenty-six her career in the air was over. Today, she is one of the five of our first nineteen licensed women pilots still alive.

Her retirement was in 1912. Seventeen years later a ticket seller at the airport in Oakland, California, approached a woman of forty-three who was gazing hungrily at the airplanes as they took off and landed. Watching her face closely, he extolled the rapture of flying into the heavens. "Really, madam, you don't know what life is all about until you've gone up."

The little woman purchased a ticket. Gravely she walked out to a biplane where the propeller was slowly ticking over. A quick, comprehensive glance over the ship and she climbed into the front cockpit.

"Now don't be afraid, madam," said the pilot. "Just relax and enjoy yourself."

Mathilde Moisant said nothing. But as she looked down at the town which they circled a little smile came to her lips. Fear? Sitting here in a cockpit cozy as a hotel room! This was not the wind-blown, perilous air which she and Harriet Quimby had tasted when they rode between the struts of the world's early airplanes.

Katherine Stinson

OF THE GREAT FLYING STINSONS

☆

PARIS WAS SPARKLING on that May afternoon in 1918. A fleece of clouds hung above the chestnut trees of the boulevards. The kiosks looked so gay that one could hardly believe they were standing still. Under that brilliant light the uniforms of khaki and horizon of blue which stippled the Café de la Paix lost their tragic meaning. Could it possibly be that these were real soldiers engaged in a real war? Or were they figures in a Viennese musical comedy?

At one of the tables a young American aviator sat alone. With a straw he tinkled the ice in his crème de menthe, and as he did so gloomy eyes strayed from the dome of the great Opéra House to the pedestrians. Now a black-robed curé. Now a beribboned *bonne* with her charges—two dandified little boys. Now a pompous old gentleman packed away in white whiskers and beard that looked for all the world like excelsior. Then at last, when he was feeling that "gay Paree" was the loneliest place in the world, another American aviator sauntered up to his table.

Friends in a moment. That was the way of World War I when two Americans met in that alien land whose cause was one with their own. Immediately they were exchanging information about themselves. At last they reached it—that one astounding common ground. They both had trained at the Stinson School near San Antonio, Texas.

"Think of it, your having gone there, too!" gasped the first young officer.

"I sure did," breathed the second. Then eagerly he leaned across the table. "Tell me—did Katherine Stinson teach you?"

"What! You, too! *Katherine Stinson!* Say, boy, everything I know about the air I got from her."

The vignette of World War I which we have given is fictional. Yet it has its roots in truth. Indeed, time and again American aviators in France found themselves linked by that one common strand. They had studied under Katherine Stinson. Some Canadian airmen shared their background. And whenever these boys met overseas they went into a huddle of gloom. What a shame the Government wouldn't accept her services as a flier! Think of a great aviator like that being wasted over here as an ambulance driver! Why, any darned old woman could drive an ambulance! Whereas, *Stinson!* . . .

Time now for a flash back. Katherine Stinson was born in Alabama on St. Valentine's Day, 1893. That made her ten years older than the airplane. She had, in fact, hardly reached her teens when a wave of air enthusiasm swept the world. From the very first stories of airmen—yes, and airwomen, too—had touched her imagination. But she was no child to let mere daydreams take the place of life. She wanted to live those stories herself. How she succeeded makes one of the most exciting and constructive chapters of woman's career in the air.

Nineteen twelve was a pivotal year in our history. The Democratic Convention of that year nominated a long-faced, professorial man named Woodrow Wilson for the Presidency. He would never have wrested power from the Republicans, so long entrenched in the national capital, had it not been for one fact. Ex-President Theodore Roosevelt, by forming the "Bull Moose" faction, split the vote. In November America rubbed its eyes. The Democrat, Woodrow Wilson, had won. In this excitement it was natural that nobody should notice a handsome, broad-shouldered young man of thirty who had been swept into the New York Senate this year. A Democratic winner in a Republican stronghold, Franklin Delano Roosevelt had stepped for the first time into the political spotlight.

More frivolous changes were recorded in this year of 1912. A world which had long been content with the waltz and the two-step now abandoned itself to a new rhythm. It was the turkey trot. Wherever you went, whether in Shamokin, Pennsylvania, or the smartest restaurant in New York City, you were sure to hear the refrain "Everybody's doing it, doing it, doing it. Doing what? The turkey trot." Be it said in passing that the ladies who undertook this new step were often conditioned. For Paris had added a new feature to the long skirts which were now the mode. It was the "hobble," a band about the ankles guaranteed to make every woman move like a frozen-toed hen or a geisha girl.

Was it in a hobble skirt that petite Katherine Stinson with the curl over one shoulder came up from the South to Chicago to interview the famous flier and teacher, Max Lilgenstrand? History often neglects such vital details as this. All we know is that, whatever her original garb, she was soon to change it for leggings, jacket, and felt-rimmed spectacles. For she became a pet pupil of Lilgenstrand, and it was no time before she got her license. She received it on July 24, 1912, a little more than three weeks after Harriet Quimby plunged to her death in Boston Harbor. Thus, at the age of nineteen, she became the fourth licensed woman pilot in the United States. Mrs. Julie Clark, the third, was killed in Springfield, Illinois, soon after her qualifying flight.

Who was the first American woman to loop the loop? Certain authorities claim precedence for Ruth Bancroft Law, our fifth licensed flier. Others, however, maintain that the title belonged rightfully to Katherine Stinson. Perhaps all are right. Perhaps the trick was turned almost simultaneously. At all events, the two attractive young women were to meet early in their flying careers, and thereafter they often appeared together. When the duet did their famous loops in a community the public was apt

to forget the Mack Sennett comedies and even that outstanding vampire of the silent screen, Theda Bara.

Early in her career Katherine turned firebird. She was,
in fact, the first exponent of feminine skywriting. One of
her most breath-taking exploits had for its background a
military, naval, and aviation tournament given on Long
Island. On that foggy night the big crowd waiting for her
exhibition suddenly held its breath. Yes, there it was at
last—the far-off whir of a motor. But where—where? Another minute and the answer flashed from the west coast
of Coney Island. That white magnesium light showed that
the Stinson plane was about to go into action.

Thousands of uplifted eyes caught the two aerial somer-
saults before the sky again turned black. But the whir continued. It grew nearer, louder. Then—a miracle of swift
phosphorescence—the plane appeared over the Sheepshead
Speedway. Two aerial spiral twists, a loop, and at last the
young Southerner was climbing out of her seat to meet the
thunders of applause which greeted her.

"The air bothered me a lot," she admitted to the crowd
surrounding her. "After I left Coney Island I didn't know
just where I was, for my fireworks had blinded me and I
was afraid I wouldn't find a good place to land. I knew
the Speedway was plenty long enough, but I wasn't sure
just where it was."

Let us proceed to 1916. President Wilson was nearing
the end of his first administration. Over two years of that

administration had hung a cloud—distant, but increasingly
ominous. In 1914 Kaiser Wilhelm of Germany had ordered
his legions into France. Europe was twisted and bleeding
on the rack of war. America, assuming a role of strict neutrality, sold arms to those combatants who commanded the
sea lanes. These happened to be the Allies and, eloquently
as we might say to Germany that her position was merely
a matter of luck, the Kaiser's men did not catch the point.
Time and again their submarines sank our merchant ships.

Punctiliously President Wilson protested each sinking.
"Wilson's writing another tut-tut note"—such was the com-
mentary of many an American during these days. Yet there
were more Americans who approved his caution. On the
slogan, "He kept us out of war," he was carried back into
office. And this time the same handsome, broad-shouldered
young man who had gone to the New York Senate four
years before entered upon the Washington scene for the
first time. Franklin Delano Roosevelt was made Assistant
Secretary of the Navy.

Before this happened Katherine Stinson had been making
an international reputation. In December of 1913 we hear
of her being in London. "Why shouldn't I fly the ocean?"
seems to have been her favorite opening in any London
drawing room. So far as we know it is the first time that
the ambitious idea had ever entered any woman's head.
And that it should have come from such a woman! Our
British cousins looked down at this wee thing with her soft
Alabama drawl and her curl over one shoulder and they

must have said, "Have a spot of tea, my dear, and you'll feel better."

Well, she might not have been able to convert the English to her transatlantic flight, but she did convince them that she was a flier. For what a lark the little Southerner did have a few days before Christmas! She sailed around over London—the Houses of Parliament and the dome of St. Peter's, the Christopher Wren steeples and the dome of St. Paul's—and then just for good measure—yes, of course, a few of her famous loops. Watching her pranks, an Englishman drawled, "I say, that's not a girl. That's Puck of the skies."

It was just about the time of Wilson's second election that she became the first woman to fly in the Orient. If anyone wished to devise a ballet he could find no better theme than that visit which carried her over Japan and China. Imagine her, for example, landing in the skies above Peking. Imagine the slanting eyes which looked upward to find that strange apparition high above the shrines of the ancient walled city. Imagine the throng, bright with mandarin coats, silken trousers, and cone-shaped coolie hats, gasping to see, not a man, but a girl in her early twenties, descending from the strange airplane.

A climax in the ballet came when the President of China, Li Yung Hung, who had been tremendously impressed by her flying, gave her his personal check for several thousand dollars and also a silver loving cup. Today the latter is one of the most cherished trophies Katherine Stinson, who has been honored by so many medals and decorations, ever received. For the cup, originally presented to "Miss Shih Lien Sun, Granddaughter of Heaven," bears the inscription, "A Thousand Li in the Twinkling of an Eye." Nor was this all. The president asked her to make a special flight for him and his guest from the grounds of the Sacred Temple at the Imperial Palace in Peking. At the end of this he presented her with a magnificent diamond pin.

However, life during these years was not all made up of temple bells and imperial gardens and other oriental extravaganza. Already she had started constructive work in World War I. She was teaching some of the pupils at the flying school which she and her brother Eddie, just a year younger than she, had started at San Antonio, Texas.

This brings us immediately to an arresting fact. Katherine Stinson was a great flier—yes. But she was only one of a family of fliers. Her two brothers and her younger sister Marjorie were endowed with that same fearlessness and that same genius for doing the right thing in the air which made Katherine herself so outstanding. Especially may this be said of Eddie Stinson, who made the first night flight between Chicago and New York and who had so many flying miles to his credit that he was often called "the Veteran."

As for Marjorie Stinson, although she never attained the fame of her elder sister, she was by no means an "also-ran." In the first place, her license, which was granted in August

1914, went to the youngest woman who had ever received it. She was then not quite eighteen. In the second place, immediately after she got that license she started to teach the art to a young man. For this reason she has become known as the "first flying schoolmarm." Shortly afterward Katherine taught her brother Eddie to fly. It was small wonder that when the Stinsons started that aviation school in Texas they did not have to look far afield for instructors. How about big sister and little sister? These two girls were to turn out some of our finest pilots, American and Canadian.

If you leaf over the files of *Aero Digest* you will find in the issue of February 1928 the diary of Marjorie Stinson when she was taking lessons at the Wright School at Dayton, Ohio, in 1914. Nothing could throw more light on the aviation school of a generation ago than either the text or the accompanying photographs. The machine in which she learned was a Wright B biplane of the type known as "pusher." This placed the two propellers behind the wings and compelled the flier to sit out in front in a little open seat with nothing but the skid brace before her.

Thus ensconced in "the undertaker's chair," the pupil grasped, not only one "joy stick," but two. This is to say that there was a shoulder-high lever at the left for elevation and another at the right for "warping the wings." Later on the latter was to be outlawed by a brilliant invention of Glenn Curtiss, great rival of the Wright brothers. It was the wing flap, or aileron control. A parallel advance was made subsequently by the invention of the foot rudder. But in the meanwhile poor little Marjorie and her companions had to cope with a hinged handle on top of the elevator lever.

Marjorie acquired her license after about four and a half hours in the air. To modern ears this has a fabulous sound. Ah, but do not forget that she had her groundwork. Her plane was at first balanced on a wooden horse and then thrown out of balance by an electric-driven motor. To get her ship back into the correct position by means of that wing-warping device—this was one of the chief lessons of the Wright School. Primitive as the device was, today's Link trainer, that last word in the practice of instrument flying, traces its ancestry straight back to 1914's mechanical plane.

Marjorie's photograph in a girly-girly summer frock accompanies other shots which show her at work on her plane. Here she is, looking very much like her elder sister —she, too, affected the curl over one shoulder—and you can hardly believe that this seventeen-year-old ever did anything more strenuous than lie in a hammock and listen to some swain strum on his mandolin "Sweet Adeline." Some of the entries in her diary corroborate this impression. She goes to a Hofbrau house but is bored, for—"I never did like beer." Someone remarks that a man is handsome and she thinks to herself, "If you could just see some of our Southern men!" Yet in the air this little Southern belle was a lion-hearted Stinson.

Nineteen seventeen! Rat-tat-tat of drums. "It's a Long, Long Way to Tipperary." Army camps springing up from the peaceful countryside. Sale of Liberty Bonds. Red Cross drives. At last the long period of "neutrality" was over. President Wilson had written his last note to Germany. The country was at war.

The Stinson School of Aviation at San Antonio, Texas, was caught up in the surge. It was enrolling more and more pupils. Both sisters took a hand in the instruction. But this did not satisfy Katherine Stinson. She, like her friendly rival Ruth Law, wanted to go overseas to take a hand in the fighting. In both cases the Government rejected these fiery Amazons.

But in the summer of 1917 Katherine performed a notable feat for her country. One of the favorite military models was a Curtiss J-N tractor plane affectionately known as the "Jenny." With a background of only five minutes' practice in this biplane, the girl tucked herself into the cockpit and started on a Red Cross drive from Buffalo to Washington.

On the way to Albany she spied below her the Empire State Express speeding to the same destination. The sight of that train inspired all the Puck in her nature. Circling overhead so as to give the passengers full benefit of the sport and flying sometimes at a height of only 300 feet above her competitor, she raced the express. What was more, she beat it by thirty-four minutes.

The last lap of her trip—that from Philadelphia to Washington—gave her no chance for her usual pranks, but it did give her a chance for a skill which was to leave modern fliers such as Amelia Earhart gasping with wonder. She flew with only a map torn from a timetable to guide her! Landing in front of the Washington Monument, she found a cheering throng of 5,000 people to greet her. They cheered still more when she handed over to the late William G. McAdoo, then Secretary of the Treasury, $2,000,000 for the Red Cross.

The New York *Sun* commented editorially upon the wonderful dexterity with which this slip of a girl handled the heavy army plane. Their tribute to her was turned into a dart against the many young men who were afraid to go into aviation. "When girls can manage an airplane," they wrote, "it should be a reproach to men who shudder at flying."

The war gave her a chance at a big "first" in woman's story of the air. She was the pioneer of her sex to carry mail. On one of these occasions she, together with the Red Cross letters and documents which she was carrying, had a narrow escape. Flying nonstop from Chicago to Binghamton, New York, her machine turned turtle when she attempted a landing at the latter city.

Incidentally, this nonstop flight of 783 miles which she achieved in 10 hours bettered by nearly 271 miles the record made by her friend Ruth Law in 1916. It also eclipsed her own nonstop trip from San Diego to San Francisco. Yet the latter, completed on December 11, 1917, made aviation history. Those 610 miles between the two California cities was the longest nonstop flight yet made by either American man or woman.

She made this California flight after a meal which Gandhi himself wouldn't have called self-indulgent. It was half a boiled egg. Starting at sunrise from San Diego, she had been able to get no attention from a waitress whose creed was obviously, "If women are fools enough to go up in the air, let 'em go empty." But let us hear Katherine Stinson's own words about her record-breaking flight.

"I left San Diego at 7.31," she writes, "and headed north through the heavy fog banked over Los Angeles. I thought first of the record I was trying to break, but once in a while thoughts of the spring fashions and that sassy waitress in San Diego came.

"Passing over Los Angeles, I began to rise gradually over Tehachapi Pass, 9,000 feet above. I knew that aviators had tried to cross it and failed and I knew, too, that once over the top I would have no trouble. I had practiced mountain flying in Alabama and it was easy.

"Beyond Tehachapi the sky cleared. The beautiful California landscape spread under me like a huge painting as I sped along at the rate of 62 miles per hour. Away down below I saw a lot of children playing about during the noon hour. They waved at me and I waved back, but I suppose they didn't see me.

"Once I traveled over a long train and I could see the engineer looking up at me from his cab. 'What a darn little fool that is,' I thought he must be saying, but I waved at him and passed his train.

"Occasionally I shifted my map mounted on rollers, so

I could handle a great length of it. It was easy to tell where I was all the time.

"Often I could see big caterpillar tractors plowing below, and my thoughts went back to the women working in the fields of Japan.

"Towns, cities, farms, hills, and mountains passed rapidly. The cold head wind blew into my plane; it cut my lips and chilled me, but I never had any fear. The main thing was speed. I carried along my knitting, but I did not have a chance to do much of it.

"I circled around the Golden Gate and found the Presidio. Tears came to my eyes as I heard the cheers of thousands of soldiers down below. They were lined up in two files and I landed between them. They rushed up and helped me out of my plane and I was mighty proud. I'll bet Ruth Law is glad a girl and not a man broke her record."

By this nonstop flight Katherine Stinson became the empress of the American skies. Indeed, in all the world there was not an aviatrix who could challenge her. Even a planet preoccupied with war and rumors of wars could take note of her record, and her fame spread from land to land. Because those hours between San Diego and San Francisco did represent one of the great historic flights of Woman we are glad that the future may share them through these words from her own pen.

It is, too, not only the subject matter which makes this document so valuable. What kind of girl was this who, a quarter of a century ago, blazed a new path for the woman aviator? In every simple, unaffected word she has written you have the answer. Feminine? She can think of the spring styles as she starts out on her momentous voyage. Compassionate? The caterpillar tractors call her back to the poor drudging women of Japan. Capable of deep feeling? Tears come to her eyes as she hears the shouts of the soldiers at the Presidio. Gay-hearted? Note her reference to the engineer and the "sassy waitress." But through all these qualities runs one supreme attribute. This girl of twenty-four knew no fear—not even of the 9,000-foot mountain which had lured so many men eagles to their death.

This California flight was one of the few which Katherine Stinson made for mere glory during our war months of 1917-18. Red Cross activities; Liberty Bond drives; teaching at the Stinson school; the knitting which she carried even on her great nonstop trip (she really doesn't fall from her pedestal when she admits she got little time for her needles on that flight)—these were all outlets for the patriotic spirit which cried, "Let me do something more to help!" However, they were not enough. And, frustrated in every attempt to fly combat, she went to France as an ambulance driver.

The nightmarish days and nights of the conflict broke that body, so intrepid and enduring in the air. She was taken to a Paris hospital and kept there for a year. She did not recover, in fact, for a long time after she came back to America. But at last the fine, light air of New Mexico brought her back to health. In 1927 the world read of her marriage to Judge Miguel Otero, Jr., of Santa Fe.

Her husband was himself an ace flier. Yet both he and his famous wife have forsworn the air. Perhaps the wings which they both so loved in their youth are stained too deep and bright with the world's blood to keep any of their old romance. At all events, their energies and talents are now absorbed by the earth. Katherine, for instance, has won distinction as a designer of domestic architecture. Not only did the Otero home for which she drew the plans win first prize in a Santa Fe competition, but she got an award for the best house costing less than $6,000.

Katherine Stinson was only nineteen when she became a licensed flier. She was only twenty-five when she gave up her career in the air. Yet those six years are packed with meaning in the history of woman's achievement. First, she possessed a genius for flying excelled by few men. Second, she was the leading spirit in America's most celebrated family of fliers, a family which was later to recruit her baby brother Jack. Most of all, however, she was the great flying schoolmarm of World War I.

It is gratifying to realize that all that this daring woman contributed to aviation during World War I is not forgotten by a generation plunged in World War II. When in May 1940 the General Federation of Women's Clubs met at Atlantic City it named Katherine Stinson as one of America's fifty-two outstanding women. It did more. Who were the three living women who had done most for aviation?

Katherine Stinson, Ruth Nichols, and Anne Lindbergh—
such was the answer of the Federation.

Mrs. Miguel Otero came from Santa Fe to Atlantic City
to receive that tribute. Curiously some of the younger
generation of aviators stared at her. Was this shy, quiet,
self-effacing woman of forty-seven the same Katherine
Stinson who had soared above perilous mountain peaks;
who had driven an almost untried military plane with the
skill denied to many men; who had raced Puckishly with
the engineers of express trains and given to dozens of pilots
the training they needed for war? Perhaps the explanation
lies in the words of another woman flier. "I hope," she said,
"that I'm a woman first and an aviator second."

Ruth Bancroft Law

MADCAP OF THE AIR

H<small>ERE</small> she comes!"

Cutting the faint drone of a motor somewhere in Massachusetts' chill November sky, the shout rang across the landing field. The waiting group of men stirred and separated. Every head tilted back to watch the biplane in the distance nose gently downward. A mechanic sauntered toward it past the motionless plane which the famous flier, Lincoln Beachey, had set down on the field an hour before. In the wake of this figure a reporter from the Boston *Transcript* strolled over to Beachey, who stood looking upward in a pose tense as a coiled spring.

"Proud of your pupil, Beachey?" asked the reporter. "She's a dandy and no mistake!"

Beachey didn't answer. He had begun to run, for the incoming plane had touched the ground and was bobbing and bumping to a stop. An official from the Aero Club, measuring tape in hand, was closely inspecting the spot where the landing wheels had stopped. The others saw only the small figure in the open seat between the struts. With gloved hands still on the controls and laced boots pressing against the footboard, the young woman fixed sparkling blue eyes upon the man in a flier's suit who reached the plane first.

"Perfect landing, Ruth!" he shouted. "You leveled her off smooth as silk!"

She laughed delightedly. The next minute she had leaped to the ground and placed both hands in those of her instructor. Congratulations buzzed over the hooded blonde head. Her response was interrupted by little pats on Beachey's arm as she insisted all the credit was his. Then the reporter, scampering off to his Model T Ford parked at the edge of the field, rushed away to write a little piece on Ruth Bancroft Law, pupil of Lincoln Beachey. Such was the successful official test which gave to the fifth American woman her pilot's license.

It was easy to make a good story about Ruth Law. In the summer of the year 1912 she had become a news feature. At the Saugus race track at Marblehead, Massachusetts, where she got her first practice in solo flying, she gave out many a good interview. Later her exploits before crowds at Providence, Rhode Island, where a woman never had flown before, aroused great enthusiasm. But this novice of twenty-five years considered mere flying only a prelude to the capers she meant to cut.

To begin with, wasn't she Rodman Law's sister? He was a well-known aviator and parachute jumper. At the Law

home in Lynn, Massachusetts, the family lived on a daily
diet of aviation facts and fancies. Second, hadn't she the
most wonderful teacher in the world? The great Beachey
had flown over Niagara Falls and had made an altitude
record of 11,742 feet. He was a cool one. With engine shut
off, he would hurtle down from the skies like a falcon
and then glide to earth in slowing spirals.

No pupil of Lincoln Beachey, gaining mastery of mechanism and the wind in daily flights, could escape one great
ambition. It was to be a sensation—a stunt flier, a madcap
of the air. And yet Ruth Law had no reason to belittle the
danger of such a program. The girl had been sitting in her
plane ready to go up for her first exhibition flight when she
saw Harriet Quimby of international fame plunge to her
death in Boston Harbor. Yet for all her sorrow over the
tragic event Ruth felt no qualms about the risks she herself
meant to take.

After all, there was Katherine Stinson, a wonderful stunt
flier who behaved in the air with all the ease of a bird.
Within a few months after getting her license Ruth herself
became a great figure in the East. In Florida she made
some nine hundred short flights and often piloted famous
passengers at $50 a trip. Soon her solo exhibitions were
supplemented by teamwork with Katherine Stinson, and
the two young fliers made several tours together. They
looped loops and zigzagged in the clouds like a pair of
happy seals tumbling in ocean waves. Although never reaching her friend's reputation as a teacher, Ruth Law did a

great deal of private coaching between her own tours. From all these efforts she was earning sums of money large enough to support the plane she had bought and cover all the many costs of flying.

Perhaps skill and energy alone could not have so rapidly built up this career; but the young woman had acquired a partner with the genius of making the most of every asset. Early in 1913 Ruth married Charles Oliver. In him she found not only a devoted husband, but a shrewd manager. Clever at business, Oliver also had a superb gift for mechanics and took charge of all the details of the plane's grooming. Reporters liked to interview Mr. Oliver because his friendliness was linked to such genuine admiration for his wife's talent that he welcomed a chance to talk about her.

"She's an instinctive flier," he told one interviewer. "She anticipates what's coming before it happens. She doesn't wait until the wind strikes her and then push her levers. She pushes them first and is ready to meet that wind."

Flying, flying, and then more flights—such was Ruth Law's record for the next three years. In the late spring of 1916 we find her at the great aviation meet held at Sheepshead Bay, New York.

In those days a racing field was very informal. Around Ruth's plane that May afternoon a sociable group stood chatting with the small blonde flier. Among them were two members of her family—her famous brother Rodman and her mother. Mrs. Law, portly and smiling, was calm, as if the two eaglets she had brought up were no more agitating than a pair of ducks. Charles Oliver, however, was nervously keen on the business at hand.

Suddenly he said, "Well, Ruth, everything's set!"

With that he helped her into her seat between the struts. She waved her hand. The motor blotted out the casual good-by on her lips, and off she went. This time her flight was to establish a new altitude record. The Laws, mother and son, continued to chat with other fliers and with each other. A long time elapsed and Charles Oliver began to pace back and forth, stare into the sky, and murmur that Ruth ought to be back. Mrs. Law and Rodman indifferently went on talking. Even when a cry went up that the flier was in sight—was, indeed, almost falling down from the sky, her two relatives preserved their unshakable tranquillity.

With a speed that left all the other spectators breathless, Ruth Law zoomed across the field and stopped. Her husband rushed to her. What was the matter? Where was her quick smile, her quick leap to the ground? With blue lips she sat staring, apparently unable to move. "I'm frozen!" she murmured.

Oliver and Rodman helped the young woman out of the plane and began massaging her hands and legs. She was shivering from head to foot. It was some time before she thawed enough to ask, "What does the barograph say?"

Charles Oliver reverted to his role as manager. He had already looked at the barograph reading and couldn't keep the disappointment from his voice as he replied, "It shows eleven hundred feet exactly. You'll get second prize. But"

—he forced a cheering tone—"you made a woman's record anyway."

Ruth sprang to her feet and her blue eyes blazed. "I didn't go out after a woman's record!" she stormed. "I wanted to beat the men. And I would have, too, if I hadn't frozen. I couldn't keep my hand on the joy stick any longer. It was come down or crash down!"

The prize for her altitude record was $250. But to a girl who could earn four times that sum for a single exhibition, the compensation was not exciting.

Interest was keen in aviation at that time because of the European war. Planes were proving of value at first for scouting and later for armed combat. Air officials from France, Italy, and England were purchasing all the planes they could get from American factories. Like every other American flier, Ruth Law watched with keen interest the development of the monoplane and of powerful twin-motor biplanes with a sixty-foot wingspread and sheltered cockpit. More than anything in the world she wanted to buy one of those modernized planes; but they were all going abroad and nobody would sell her one.

"Perhaps they look at slim little you and think you couldn't fly one of those big things!" laughed her husband.

Ruth grinned. "Well, I'll have to show what I can do in my old crate," she said. "Having flunked that altitude record, I've just got to try for something else."

She knew quite well what it was she wanted to attempt. It was a cross-country distance flight. Early in November

1916 the nation had rung with praise of Victor Carlstrom, who had tried a nonstop trip from Chicago to New York. Although a broken fuel line had forced him down at Erie, Pennsylvania, Carlstrom established a record of 462 miles. Ruth determined to do better.

The Olivers were now living in Chicago and Ruth used Grant Park for her runway on practice flights. There was always a crowd to watch her come and go. But not on the morning of November 19. If early risers made out in the half-light the group of people about the hangar, it would never have occurred to them that a flight was contemplated. Not in that gale! Nobody would be so foolish.

That was, indeed, exactly what the mechanics were saying. So was Mr. Stevens, official representative of the Aero Club who was there to record proceedings. Charles Oliver, however, concentrated only on the motor. It refused to start in the cold. Looking unlike her slim self in her two leather coats over two layers of wool, Ruth Law watched anxiously. More than an hour went by.

"Give it up for today, Miss Law!" begged the chief mechanic.

She shook her head stubbornly. Proudly she pointed out to the official all the new improvements in the plane—a little tin dashboard to protect her feet from the wind, gasoline tanks enlarged to a capacity of fifty-three gallons, and a rubber fuel line which could not break. For instruments she had a barograph, a clock, a compass, and a tachometer. It was slim equipment for the lonely venture.

This was long before any markers existed to guide the pilot and more than a decade too soon for radio installations. No airport or landing field had ever been constructed. In 1916 cross-country flying was like exploring unknown regions, and few there were who had ever dared it. Ruth had never been farther away from the take-off field than twenty-five miles.

By the time the sun was up behind the murky clouds the engine was roaring nicely. At last the whistle of the wind became a little less shrill. Suddenly Ruth jumped into the plane, made a trial spin, landed again, and cried, "All right, boys, I'm off!"

The mechanic uttered hoarse and panic-stricken cries of protest. But Ruth caught the confidence in her husband's proud glance. He and Stevens waved their hats. She gave them a last smile, then roared away, battling against the wind for a chance to rise in the air. It was just eight twenty-five.

Hour after hour she flew. Her only map was a scribble of directions on her cuff. The wind jabbed at the patched places on the wings and numbed the small gloved hands on the controls. By two o'clock she thought she must be nearing Hornell, New York. The engine sputtered. She was getting short of gas, but she had meant to stop for it at Hornell. Again that warning sputter. Now she could see the town two miles ahead. Could she make it? Not she. There wasn't a drop of gas to keep her motor going. Gliding down, she saw a race track below. It was tricky business

Ruth Law, who broke all cross-country flying records in 1916 in the machine in which she made the flight from Chicago to Governors Island, New York.

The madcap stunt flier racing an automobile.

Brown Brothers

Brown Brothers

Katherine Stinson, member of a famous flying family, a noted teacher of aviators for World War I, and first woman to fly the mail.

Brown Brothers

to reach it without power and in the ceaseless wind. A crowd of people surrounded the green oval. "Charles must have wired about my flight!" she thought.

Then as the wheels touched earth and she cut off the motor, a glance at her watch told her the stupendous fact. She had broken Carlstrom's record! In 5 hours and 45 minutes she had covered 590 miles in a nonstop flight.

Waving to the crowd, she shouted the news! They cheered and cried her name to the sullen skies. They surged about the plane and lifted her down. What could they do for her? She was shivering. She needed food and warmth.

Laughing, she pushed them off. "Help me tie my plane to that tree!" she cried. "It might blow over in this gale!"

When that was done, she allowed herself to be escorted to a waiting motorcar and whisked to Hornell for lunch. The whole town was seething with excitement. Wires were flashing the news over the entire country that a new national distance record had been made by a woman. Reporters from the local paper pressed about her. Yet for all their enthusiasm, the citizens didn't know the full triumph of that venture. Only a flier could appreciate the feat of Ruth Law, utterly inexperienced in cross-country flying, piloting that antiquated little crate through the second longest flight in the world.

Ruth said to herself that Charles Oliver would be triumphant. As she winged her way on to Binghamton that afternoon, she felt she could almost hear his words of praise. She went no farther that night, but was almost too happy to sleep. Next morning messages poured in upon her. She would be welcomed officially at Governors Island, she was told. Cheered to the echo by all Binghamton, she winged her way down to New York.

Skimming over the tall buildings, Ruth heard again the threatening cough from the engine. For a few moments she wondered whether she would have to land in the city streets. But, escorted by a flock of hawks, she finally glided down on the island in the harbor and stopped near a huge battle plane. It belonged to no less a person than the man she had just bested, Victor Carlstrom. A battalion of photographers and a large crowd were gathered to welcome her. She saw General Leonard Wood step forward to greet her. Eager hands unfastened the strap that held her to her seat, and a cheer went up that must have reached Manhattan.

General Wood seized her hand. Beaming, he said, "Little girl, you beat them all."

Ruth thanked him. Then a wave of weariness went over her. "I'm so cold," she said piteously.

Mercifully she was swept away from the crowd to a welcoming residence on the island. There, warmed and refreshed, she was soon ready to be lionized at the luncheon party prepared for her. But before being presented to the guests, she said to her hostess, "Has anyone a powder puff?" The utterly feminine question from this daring victor of the air delighted all the women who were present.

New York claimed Ruth Law for weeks. In return for

the royal welcome the city gave her, she put on a performance one night that people remembered for years. With white magnesium streamers of light floating behind her plane, she flew across the bay and circled the Statue of Liberty. Then she looped the loop and traced the word Liberty across the sky.

That story made the first page of the New York *World*. Male fliers of long experience told Miss Law they would hardly have had the courage for such a stunt. She was guest of honor at a dinner given by the Civic Forum and the Aero Club. Sitting between Rear Admiral Robert E. Peary and Roald Amundsen, Ruth realized with the sense of a fairy tale come true that she was now a public character, an aviatrix of national fame.

Other testimonials to her sudden celebrity, however, were not so delightful. With a rush the commercial advertising world was upon her. In her hotel room the telephone rang day and night. Would Miss Law consent to be photographed wearing a certain petticoat or sports suit, using a particular brand of soap, or nibbling a well-known type of breakfast food? Dozens of firms made her financially tempting offers. She was wanted for lectures and personal appearances on the stage. Sometimes Miss Law was amused and sometimes indignant.

"What do you think I've been offered today?" she asked a friend mirthfully. "I could earn $35,000 by riding upside down on a motorcycle in a Broadway show!" Suddenly her blue eyes flashed. "A pilot is not supposed to have any human dignity. We're just circus people still."

One proposal was rather tempting. It was to manage an airplane factory at a salary of $500 a week. In the end, fearing the company wanted to use her name only as a bait for the sale of stock, she refused the position.

What Ruth did want was something for aviation. One afternoon at her suite in the McAlpin Hotel she invited a number of Aero Club officials, pilots, and people seriously interested in the future of flying, together with reporters, to discuss a subject dear to her heart. Cities and towns should be so marked, she said, that they could be identified from the air. As she sat talking, with blonde hair artfully arranged, clad in a georgette blouse and velvet skirt, Miss Law amazed those cub reporters who had never seen a woman flier before. One of them said, "And I thought she'd be covered with oil and breathing out gasoline!"

The respect with which both aeronautical officials and the press treated Ruth's plea for air-marking towns added more luster to her name. Journalists thought she knew what she was about. Knowledge and demonstrated skill plus the fact that she had always been good "copy" inspired the New York *World* to make Ruth Law a sporting offer. That newspaper was one of many which was preaching the necessity of America's preparing for war. Continued sinkings of American shipping by German submarines, the stalemate of the Allies' fighting in the trenches of Northern France, the tight-drawn lines of sympathy between this country and Great Britain—here were elements which pressed for action. Magazines and newspapers were playing up heroic stories of aviators and the increasing use of planes in combat. The *World* decided it would be at once valuable and emotionally appealing to readers to have Ruth report from Europe news about aviation at the front. Definite information concerning the performance of various types of planes would help the American Air Force. Seeing the war through a woman's eyes would be interesting to the public.

Early in January 1917 Ruth Law and her husband set sail for Europe. They visited plants and training fields in England and in France. Although they were allowed no nearer the front than Compiègne, where the German army was first turned back, they gathered a store of facts and impressions. Ruth met the great romantic French ace, Georges Guynemer, who gave her a ring made from a button of a German flier's coat. She watched young British and French aviators start off from behind the lines and studied the various makes of planes.

Climax of the trip was a flight from the vast air center Le Bourget near Paris. One of the French plane designers took her up in his new model. At a speed of 198 miles an hour she was flown over Paris. That was thrilling enough for the average woman, but this particular passenger was more interested in the craft they passed in the air than in looking down upon the Eiffel Tower. Dirigibles, scout planes, and fighting planes sped along the air lanes.

On the way back the Frenchman said, "Now, I will put the ship through its paces!"

Of course he was showing the greatest possible courtesy to a visiting pilot of distinction. Nevertheless, the experience was hair-raising. To the readers of the New York *World* Ruth described it thus:

First we stood up on one wing, then with the ease and grace of a bird we dived and spiraled and then climbed again. Then in beautiful, great circles we came floating down past rows and rows of military hangars, coming to a stop in front of our waiting automobile.

On April 4 the Olivers landed once more in New York. In her huge felt hat, accompanied by a big Belgian police dog named Poilu, Ruth was the focus of attention from reporters who met the ship. From her very first word it was evident that she had returned fired with zeal for pushing aviation in the United States. Eagerly she described the modern fighting planes she had seen. "France is ten years ahead of other countries!" she cried.

Her most sensational statement, however, was that she herself intended to offer her services to her country. What services? Well, teaching, of course, and any other task which might be useful. Suddenly the small flier flung back her head and cried, "I could drive a machine with a gun and gunner and go into actual battle. That's what I'd like to do more than anything—get right into the fight."

The United States was now on the very eve of a declaration of war against Germany. When that fateful declaration was made, Ruth Law was enlisted by the Red Cross at once. All through the summer and fall of that year she was flying, speaking, and giving exhibitions for the benefit of the Red Cross and the sale of Liberty Bonds. North, South, East, and West she dashed about the skies. She dropped "Loan Bombs" over Cleveland, gave war talks in Illinois, and performed her most daring air stunts in New Orleans.

Once at the national capital, in the interests of the Red Cross, she flew straight down Pennsylvania Avenue, skimming between the buildings. "I investigated first," she said afterward, "and found there were no wires in the way. It was perfectly safe."

In November, however, she presented her personal petition to the War Department to be enlisted in the Air Force. Secretary of War Newton D. Baker was a warm supporter of feminism, but he stopped short at granting the doughty little aviator a commission. All she could do was to go on with her activities on land and in the air for the support of war effort.

Not till after Armistice Day did she return to her personal career. But that comeback was prompt. In November 1918 she flew from New York City to Chicago in 8 hours and 55 minutes, which was noteworthy speed for that era. The next year she and her husband were in the Philippines. There an air-mail service was just in process of organization. Miss Law pushed it several points ahead by flying the mail herself a number of times. For two more years she continued to earn large sums of money as an exhibition flier. But at last she was involved in a stunt which ended in death for a fellow madcap.

A girl by the name of Madeline Davis was trying an open-air circus stunt. This consisted of jumping from a swift motorcar to the rope ladder of a plane as it passed over the automobile. Ruth Law was the driver of the speeding car. The girl caught the rope, but hadn't the strength to pull herself up, and fell to her death.

Doubtless this tragedy put the last strain on the over-wrought nerves of Charles Oliver. For many years he had been on edge with anxiety about the wife he loved. When at last he collapsed Ruth declared immediately that she would give up her career. As Mrs. Charles Oliver she went with him to live in Beverly Hills, California, and only once again did she emerge from happy privacy as Ruth Law.

That was in 1927 when Commander Byrd, Captain René Fonck, the famous French ace, and several other pilots had entered the transatlantic flight competition for the $25,000 prize. Interest in the great contest was at high pitch all over the country. Suddenly a new note was injected into the news. Ruth Law was in the East and in action again. She was planning to be the first of her sex to fly across the ocean. Stories of her record-breaking career once more appeared in the journals. It was she, stunt flier of 1912, and not an unknown young man named Charles Lindbergh, who was touted as a rival for the celebrated pilots.

History, however, is not written by the prophets. Lindbergh sped off to glory. And Ruth Law Oliver returned to California.

Not long ago a magazine writer interviewed the famous flier of another day. Mrs. Oliver talked of aviation with no apparent nostalgia for the career she had given up. When she was asked point-blank if she wouldn't like to fly again, she laughed.

"Oh well, things are so proper now. A pilot has so many rules and regulations to follow, it wouldn't be much fun. I couldn't skim over roof tops today or land in the streets or on a race track. The good old crazy days are gone."

Thus from her disguise as a dignified matron of Beverly Hills peeked out for an instant the "Madcap of the Air."

Phoebe Fairgrave Omlie

HEROINE AND EXPERT

☆

I UNDERSTAND you have a lot of army planes for sale and I'm here to buy one."

The low voice was very matter of fact. For an instant the Curtiss Company demonstrator looked at the speaker with eyes growing round and jaw almost unhinging itself. "*You?*" he gasped.

Before him in the Curtiss Field office stood a small person in ankle-length sports suit and sailor hat. Her bobbed brown hair framed a round little face which managed to look both pert and earnest. Her brown eyes sparkled with light. She hardly looked her eighteen years.

Trying not to laugh, the salesman asked the girl to sit down and talk over the proposition. Swiftly he drew from her certain pertinent facts. Phoebe Fairgrave, born in Des Moines, Iowa, now lived with her parents and brother in St. Paul, Minnesota. She had taken courses in mechanics at high school but didn't know quite why until President Woodrow Wilson on his postwar League of Nations tour came zooming over the school building in his plane. All at once she knew she was going to be a flier. She'd tried other things—had just left a good job with an insurance company. But she couldn't be cooped up. Taking the $4,000 her grandfather had left her, she had come on to Curtiss Field to buy a plane and take lessons. When could she begin?

All this was too good for one man to enjoy by himself. The flier summoned others and presented them. Ringed around by company officers, the girl looked happily expectant. One of the men asked, "Miss Fairgrave, have you ever been in a plane?"

"No," she confessed; "but I'm dying to go up."

When the flight was arranged, the girl was radiant. She didn't know that the pilot had had traitorous instructions from his superior. "Give this kid the works, Ray," the officer had said; "loops, barrel rolls—everything. Make her good and sick. She's got a yen to be a flier and the notion must be squelched quick."

Clever as it was, the plot didn't work. Now and then during the wild ride the girl grew pale and gritted her teeth, but when she emerged from the plane on the landing field, she cried, "It was marvelous. Now when can I start my flying lessons?"

"You're too little and much too young to learn to fly!" That was what the Curtiss pilots all responded. It was enough to daunt the hardiest spirit. Then a slender, serious-looking fellow came to her rescue. He was Captain Vernon

Omlie, a skilled bombing instructor for the Army during the war and now, when not engaged in stunt flying, a teacher for the Curtiss Company. Gravely he listened to the girl's plea. Then with a gallantry which defied the caution of the others he said, "I'll teach you to fly. I'm sure you have the nerve for it."

Between lessons the two talked about the future of aviation. This was in the spring of 1921, and the era of prosperity was still in the distance. Captain Omlie thought immediate prospects were black. Interest in flying had petered out since the war. Capital was hard to get for production of planes or promotion of travel.

"America hasn't yet the faintest idea what aviation is destined to become," said he. "People think flying is just another sensation and that fliers are merely acrobats. There are no solid opportunities for a man pilot now, let alone a girl. I earn my living by stunting. Without an income, how are you going to exist?"

Perhaps it was in answer to the question that Phoebe first got her idea. She voiced it after a bit in that perky way she had. "I'm going to do stunts, too. I'll learn parachute jumping."

The circulation of that announcement about Curtiss Field produced horrified protest. Not a single pilot would treat the idea with anything but scorn. At last, however, the chivalrous Omlie again reluctantly took on the role of destiny. "If you're crazy enough to try this," he growled, "I'll take you up for a first lesson."

Naturally he couldn't help hoping that one lesson would forever quench the mad ambition. With a harness fastened about her and a rope attaching her firmly to the plane, Phoebe was to climb over the cockpit and crawl out on the wing. When the captain had risen about 2,000 feet, he told her to go ahead.

Taking a deep breath, she started. To her amazement the wing was steady as a floor. The strong wind didn't blow her off, but pinned her down tight. She was neither dizzy nor afraid. Calmly looking about and making tentative moves along the wing, she was more sure than ever that she would have the courage to make the jump. Several times she went up with harness and safety rope. Then came a trip when Captain Omlie forgot to make the rope fast. Phoebe always had the parachute strapped on and ready. When she got out on the wing that day and found herself free, she made up her mind at once. "I'll do it now!" she thought.

At the far edge of the wing she looked back. White-faced and aghast, the pilot was motioning her to return. She shook her head and began to make her careful way to the undercarriage. She had folded the parachute herself. Had she done it right? "I'll soon find out," she thought with grim humor. Head downward, she hung for a moment. A long way to the ground it looked. Her heart beat hard. She let go.

In the rush of falling she almost forgot to pull the cord.

PHOEBE FAIRGRAVE OMLIE, private flying specialist for the Civil Aeronautics Authority.

MRS. OMLIE, right, and BLANCHE NOYES, at the time of the 1930 Air Races in which they both took part.

When she did so, the opened parachute jerked her upward. Relief and strangeness blurred her sensations until suddenly she felt herself brushing into the top of a tree. She had missed the landing field entirely. It was not hard to climb down from the perch. When she reached the ground she looked up into the sky and uttered a silent shout of triumph. She had done it! And it was no worse than coming down in a fast elevator.

Young Captain Omlie was more than vastly relieved. He was proud. "You're all right, Phoebe Fairgrave!" he cried. "You're slated for the flying circus."

Practicing the new skill meant a series of adventures. Once Phoebe came down in a lake. She couldn't swim, but the parachute kept her afloat and slowly drew her to shore. When she grew quite accustomed to the jump, she decided to try a double parachute trick. After a short descent, she cut away the first parachute and then pulled the cord of the second. Between the two operations lie seconds of unequaled suspense.

Phoebe had been writing her brother about her exploits. He was not only sympathetic, but so envious that he determined to follow her lead. The upshot of his response and of Omlie's interest was that all three of them resolved to join the Flying Circus which a famous pilot named Glenn Messer was operating at Des Moines. Before the trio arrived, however, Phoebe accomplished a feat which made her a magnificent asset for any show. She returned to St. Paul that summer, and one afternoon, with practically the whole city gathered to watch, this eighteen-year-old made a parachute jump of 15,000 feet. This was a world record for women.

Of course a circus has to offer plenty of thrills. The three main performers of Glenn Messer's outfit practiced for hours in an old barn before trying out stunts in the air. One of the features of the show required almost incredible skill and daring. Phoebe would stand on top of one plane and Messer would swing from a trapeze fastened to another. After the two planes swept close together, Phoebe would be seen dangling from Messer's hands. That she was no bigger than a child made such a maneuver possible. The girl performed acrobatics on the wing of a plane in midflight, took triple parachute jumps, and hopped from the wing of one plane to the wing of another as it passed by. It wasn't long before the motion-picture scouts discovered her. From time to time she was engaged to double for stars who had to make hairbreadth escapes by plane.

Her only bad accident during this extraordinary period was a parachute descent into high-tension wires. Miraculously escaping with her life, she received terrible burns. She knew that this accident, like all others happening to fliers, would be headlined in the paper. First she sent a reassuring wire to her mother. Then she backed up her message by going out with one arm and hand bandaged and doing stunts in the air as usual. She sent her mother the clippings from the paper reporting her return to work and then went back to bed for a fortnight.

Marriage between intrepid little Phoebe and gallant Captain Vernon was inevitable. Inevitable also was their escape at the first possible moment from the air circus to serious aviation. For a time Phoebe Fairgrave had had her own flying circus and had found it very profitable. Combining her capital with that of her husband, she joined him in renting land near Memphis. There they established an airfield and flying school. Of course Phoebe herself had gradually become a very skillful pilot.

By 1926 interest in aviation had risen tremendously. Passage of the Kelly Air Mail Bill in 1925 opened air routes to private enterprise. Passenger travel was encouraged. Night flying had proved feasible, and nearly 24,000 miles of air routes were equipped with beacons and lights. Meanwhile, plane design was changing in the direction of higher speed and greater safety. Lighter and more powerful engines were built. Propellers and radiators were growing more efficient each year. What with new opportunities in commercial service, extended interest among private persons, and increasing need for test pilots at airplane factories, there was a real demand for adequate instruction.

These were the boom years of President Coolidge. The school and flying field established by the Omlies prospered. It was hard work, but they loved it, and never was a more congenial pair. Phoebe's aptitude for mechanics made her invaluable as an aid in testing, adjusting, and selling planes. One reason why she and Vernon were good salesmen for planes and excellent teachers was because they put safety in aviation before every other goal and did everything they could to promote it.

All the more ironic was the accident which befell Mrs. Omlie one day. She was taking up a strapping young fellow for a lesson. Suddenly he "froze" at the controls. In other words, he became panic-stricken, forgot what to do, and was too confused to follow her directions. Because he was so much more physically powerful than the small woman beside him, she couldn't control him. The plane went into a spin and crashed. To those near the scene it seemed unbelievable that the two escaped with only a few broken bones.

Mrs. Omlie always says that it was the terrible Mississippi flood of 1927 which aroused that whole center of the country to the value of their mid-South airport. As soon as they got news of the disaster, the Omlies set off in their planes, and for many days and nights they worked unceasingly at rescue. With inspectors and photographers they patrolled the river to report any breaks in the levees. They carried food and medical supplies to isolated towns and villages. They transported Red Cross nurses, rented planes to news agencies and newsreel companies, and maintained a mail service between Memphis and Little Rock.

When the emergency was over Phoebe had to submit to many interviews. "Not a single one of our mid-South pilots had an accident!" she cried. "And yet we were flying over water all the time in land planes. I got more thrill out of that than from any of my parachute jumps."

Praise for their work was showered upon the Omlies from all directions. Phoebe was invited to join that world-renowned organization, the Ligue Internationale des Aviateurs. She was the first woman ever to receive the honor.

Almost immediately Mrs. Omlie won another distinction. On June 30, 1927, she was issued a transport pilot's license from the U.S. Department of Commerce which had taken over the responsibility of licensing commercial pilots. It is a severe test with a prerequisite of 200 hours of solo flying and much technical knowledge of mechanics. Phoebe was the first to pass the test.

Doubtless possession of the license influenced the Mono Aircraft Company of Moline, Illinois, to engage Mrs. Omlie as assistant to the president. She held this consultant's position for three years. Meanwhile, she began to enter the field of competitive flying which was rapidly developing. She felt that this was both a professional necessity to anyone earning a living as a flier and also that such meets encouraged the women of the country to get over their distrust of planes as a means of travel.

Phoebe's first undertaking was to join the 1928 National Reliability Air Tour for the Edsel Ford trophy. It started from the airport at Dearborn, and a huge crowd had gathered to see the twenty-four planes set off. The 6,000-mile tour was to cover thirty-two cities in fifteen states over regions rough enough to test the stability of any plane. In a tiny little plane with a wingspread of only thirty feet sat Phoebe all alone. She took with her neither copilot nor mechanic and was the only woman to compete. Reporters, who made much of her at the take-off, eagerly followed her progress in the strenuous weeks of travel. From St. Louis, San Antonio, from the perilous Rocky Mountain district and the Mojave Desert where the temperature was 110 degrees, from Seattle and at last from points toward Chicago again, the newspaper stories always carried the line, "Woman pilot and tiny monoplane still in the race."

At Marfa, Texas, after landing safely, Phoebe's plane was overturned on the ground by a fierce gust of wind. The other pilots, finding a wheel on the landing gear broken and the stocky little pilot somewhat shaken by the accident, said to her, "Quitting, Phoebe?"

"Not on your life!" she cried. "I'm going through."

And so she did. After most of the eagle-like planes had reached the Dearborn field once more, a little orange-and-black speck appeared on the horizon. When the plane glided down to a stop, out of it stepped a young woman in white knickers and white silk shirt. "Here I am," she called to the officials. "Fooled you, didn't we?" Even the winner of the trophy got no such applause as this spunky solo flier. She won the admiration of the crowd.

Next year she added to her glory. At Moline, Illinois, under the auspices of the National Aeronautical Association, Phoebe Omlie made an altitude record for women of 25,400 feet. This achievement to crown her other daring feats made experts declare that she was in the first rank of the women fliers in the world.

Such personal triumphs pointed the way to deal with the drastic economic problems now facing the Omlies. When the financial crash of 1929 struck the nation, many aviation enterprises were shattered. Rich people ruined by collapse in stock values hardly had money enough to take air-line passage let alone purchase planes. The mid-South flying field had to be given up. Captain Omlie took a pilot's job and, in addition to her work with the Mono Aircraft Company, his wife decided to enter every possible competition.

In the Dixie Derby of 1930 Phoebe crossed the line first and won a purse of $2,000. In the 1931 Handicap Race, from Santa Monica to Cleveland, she beat fifty-two entrants, thirty-six of whom were men. This grand sweepstake netted the winner $3,000 and an expensive automobile. Then she topped off the triumph by taking a first for her class of planes in a speed test for women on a thirty-mile course.

It was on the occasion of these national air races that Mrs. Omlie took the press to task. She said it wasn't fair—to men. "The trouble with a mixed race," she said to the reporters, "is that women get all the breaks—the publicity and credit for every little thing they do. Men deserve every bit as much praise and attention. But you reporters don't give it to them."

In vain the newspapermen laughed and applauded.

Phoebe hadn't finished her lecture. "What's more," she went on with brown eyes flashing indignantly, "you never give a kind word to the boys on the ground. It's they who keep the motors going. My mechanic, Ollie Walker, has seen me through two races with the same plane and engine. Often he works all night to get everything shipshape. Why don't you ever speak of these heroes?"

The protest, quoted far and wide, made the flier more admired than ever. When during election, in 1932, the young woman flew about the country to campaign for Franklin D. Roosevelt, she always had a responsive press and an audience glad to welcome her. In the South this was doubly true, for there she was not only her deeply respected self, but wife of Captain Vernon, the widely known and much-liked pilot who had taught hundreds of people to fly and had carried thousands of passengers. It was a great team, that of the "Flying Omlies."

They were so completely partners in everything that it was almost a shock when the U.S. Government made Phoebe its first woman appointee in aviation. Acceptance meant that she and her husband would have to be parted more often than ever, yet they had both used all their influence to stimulate federal activity in aviation. Mrs. Omlie had made a mighty effort to convince federal officials that they should initiate a plan to air-mark all important cities. She staged a campaign in which she invited Louise Thaden, Helen Richey, Blanche Noyes, and Helen McCloskey to take part. The latter flier was a transport pilot licensed in Pittsburgh and, as a member of the Ninety Nines and contestant at many meets, had won considerable distinction. Mrs. Omlie's choice of women fliers was obviously approved by the authorities, for when the Bureau of Air Commerce was organized all four of them were appointed in succession.

It was not to the Department of Commerce, however, that Phoebe herself was called. Her title was long enough to wrap twice around her small person. Special Assistant for Air Intelligence on the National Advisory Committee for Aeronautics—that's what she was. So off went the doughty flier to settle down in Washington. From the first the work fascinated her because it drew upon her entire equipment of knowledge, skill, and experience. Sometimes she worked in the laboratory on construction problems and sometimes at research.

When the now-famous radio broadcaster, Mary Margaret McBride, was writing syndicated articles, she found Phoebe Omlie an ever-interesting subject. At the close of 1934 the writer made her own selection of the twelve women who were then contributing most to American life in various fields. One of the twelve was Mrs. Omlie, whose work to advance aviation in the United States was proving so valuable.

To her interviewer Phoebe said, "Of course I still fly. But I'm out of racing for good and all. I never did it for fun. My present work is what I love, and it's a big satisfaction to serve an administration which is really concerned about the future of aviation. Through loans it has saved some 8,000 independent operators."

Her enthusiasm for President Roosevelt's accomplishment in this realm impelled her to campaign for him again in 1936. She flew thousands of miles on a tour of fourteen states in his behalf. In one of her signed articles to the press she spoke eloquently of all those changes for the better in the country which could actually be seen from the air—smoke pouring from factory chimneys, activity in the streets, the absence of bread lines, more travel by land and air, improvement in airports and in air-marking. She herself had quietly done so much effective work for flying operations that when Mrs. Roosevelt chose a group of outstanding American women for special praise, Phoebe Omlie was inevitably among them.

From the time she hung head down from the struts of a plane ready for her first parachute jump, this flier has proved herself a brave woman, but never was her courage so tested as on one August night in 1936. She had been very happy that afternoon, for Captain Vernon had called her up long-distance. His work was far from Washington, and every moment the two could spend together was a honeymoon. In fact, their next meeting was to be a postponed celebration of their fifteenth wedding anniversary. "I'll be up early next week, darling," said Vernon in his warm voice.

Precious memory to hold forever—that's what those words were destined to be. Later that night came news that changed Phoebe Omlie's whole world. A plane belonging to the Chicago and Southern Airline had crashed just outside of St. Louis. Everyone in it was killed—the two pilots and six passengers. One of those passengers was Captain Vernon Omlie, the most distinguished flier in the South.

Into the wild whirl of Phoebe's thoughts leaped the fear that somehow her husband might have been asked to fly the plane. Perhaps he was responsible for the tragedy. "Was Vernon at the controls?" she gasped. When she heard that he was not, the anguish in her face lessened. Next day she left for Memphis. Blanche Noyes, whose own husband had been killed in a crash not long before, saw her off. The two women shared a wordless grief with the self-discipline characteristic of dauntless fliers.

Characteristic, also, was the unflagging devotion to aviation which carried Mrs. Omlie onward. She resigned from her position with the Government only to preserve the excellent traditions of work for air service which her husband had established in Tennessee. For several years she assisted W. Percy MacDonald, State Chairman of the Aviation Commission, in formulating a novel piece of legislation. Its passage meant that Tennessee was the first state in the country to make public funds available for training aviators. When she had helped organize courses of instruction under the direction of the Vocational Division of the Memphis School Board, Mrs. Omlie returned once more to her former occupation of teaching.

It is gratifying to all who know of her work that she recently received a special honor from all the educational groups in the country. At a great national meeting held in Florida in 1942 Phoebe Omlie was given a citation for her valuable contribution to American education.

She was no longer working in Tennessee when that tribute was bestowed, for Uncle Sam had never really taken his eye off her. In 1941, when the United States began to mobilize its mighty resources against Fascism, the Government called this outstanding expert back to Washington. Her new title was Private Flying Specialist for the Civil Aeronautics Authority. In that capacity she visited more than two hundred airports throughout the country to select those best suited for the training of ground crews.

In fact, the Government can claim no more valuable public servant than this widely experienced aviator. To others of her sex Phoebe Fairgrave Omlie flashes a beacon light along the airways of the future, for she is a woman who has risked death a thousand times, who has known how to love, how to fly, how to teach, and how to serve her country.

Ruth Nichols

AVIATOR WITH A PURPOSE

☆

ATLANTIC CITY in the summer of 1919 was gay as a beach umbrella. World War I was over. Its aftermath of national problems—unemployment and the failure of Prohibition—had not yet darkened the jubilance of victory and the return of the soldiers. Up and down the Boardwalk women in linen suits or frilly frocks nibbled salt-water taffy or dropped into the movies to feast upon the beauty of Lou Tellegen.

Amid the colorful crowd one afternoon a group of people made swift way to a waiting car. Dominating the others packed into the vehicle was an aristocratic-looking woman in an ankle-length ruffled frock. Tucked in beside her, a slim girl in sports clothes leaned forward eagerly to the older man opposite whose military bearing and athletic look belied the tailored suggestion of a successful member of the New York Stock Exchange.

"Dad," said the girl, "you don't think he'll go away before we get there, do you?"

"He won't," answered the man, and flashed her a smile filled with an impish look she knew so well. It always denoted some plan for fun withheld from the likely disapproval of the lady opposite.

The latter now turned to ask, "Ruth, who is this flier you are so bent on seeing?"

"Why, Mother!" The piquant face with its deep-set gray eyes was slight with enthusiasm. "He's Eddie Stinson, our number-one American flier of the Army. His sister Katherine is a great flier, too! I want to go close and get a look at his plane."

Behind the great hotels and cottages of Atlantic City was an open stretch of land near the inlet. Far as it was from the piers and the shops, Eddie Stinson had created there one of the features of the resort. American interest in flying had been given a new impulse that spring by the great feat of the Nc-4 Navy flying boat which was flown across the Atlantic by Commander Jack Towers and Holden C. Richardson. Consequently, there were always intrepid passengers ready to pay ten dollars for a brief trip in Stinson's plane. That afternoon a considerable crowd stood or sat on the benches at the edge of the open field to watch the biplane refuel with gasoline or take on passengers. Leaving the others in the party on the sightseeing benches, the man and the girl hurried down to the plane.

Probably today nobody would dignify Stinson's worn and shabby contraption by that term. Its frail construction

justified its colloquial name of "crate." A tall, weather-
beaten fellow in an ordinary suit with goggles pushed over
his forehead stepped forward to greet the newcomers. The
deeply grooved lines about his eyes and mouth suddenly
vanished in a radiantly attractive smile.

"What can I do for you?" he drawled in the soft accent
of the South.

Then the man of impeccable stockbroker attire again
turned that impish look upon his companion. "I thought,
Mr. Stinson, that if you're sure you'll bring her down in
one piece, I'd let you take my daughter up. She's been beg-
ging to go and—well, this is a graduation present to Miss
Ruth Nichols."

Anticipation mixed with glee filled those wide-apart
gray eyes. "Wonderful!" she exclaimed. Then her eyes
traveled back to the spectators' bench. "Of course you
didn't tell Mother!" She hoped that from such a distance
she could not be seen, but noticed uneasily that Stinson's
mechanic had followed her glance with a knowing grin.

The next moment Ruth was being helped into a seat in
the open cockpit of the little plane just behind the pilot.
Starting with an explosive din, the motor stifled the "ohs"
and "ahs" of the crowd. Down the hard sand raced the
wheels, and suddenly Ruth Nichols was aware that the
plane had left the ground. She was flying!

Blue water and the curving foam along the beaches, roof
tops patterned like a Cubist painting, the piers mere frail
white pencils thrust into the sea! It was beautiful. Up and

up the plane climbed, until the earth was just a map spread out below. But, oh, what was happening now? The plane was diving toward the earth, and the girl felt her head forced down on her chest. Out of the corner of frightened eyes she saw she was suspended upside down. Banged flat in her seat, with choking breath, the girl felt her neck almost snapped in two. As pressure on it lessened, she realized in a flash that Eddie Stinson was looping the loop. But even as she understood, the plane began another downward rush, and in a fury she thought the terrifying maneuver was being repeated. Then once more the plane was smoothly circling in the dazzling light.

The pilot pulled back the throttle for a glance around at his passenger. White-faced and grim, she screamed, "Why did you do that twice? Once was too much."

He grinned. "Only once," he yelled back. "Just completing the loop. Thought you'd like it."

She hadn't liked it. Still less had she liked being frightened. Why had she been? she wondered. Probably because every phase of the experience was so new and so different from any previous conception. If she could once pierce the mystery of flight and understand the manipulation of a plane, she probably wouldn't be so scared. Hardly could she enjoy the rest of the trip or follow the gliding movement of descent for this analysis of her emotions. Ruth Nichols recognized a challenge when it was offered, and she could never rest until she took it up.

The very instant the motor stopped she began to ask

Eddie Stinson questions. How did he make that loop? What was their speed? When did he know the right moment to nose downward? So intent was she on his demonstration that she paid no heed to people pushing about her or to her father's insistent question, how she had enjoyed the ride. Only as she started to leave the plane was she aware of what the laughing mechanic was saying.

"Miss Nichols, your mother is in a terrible state. I ran over to her and said, 'Look at your daughter looping the loop!' She didn't even know you were going to fly and nearly fainted."

"Oh, my," groaned Mr. Nichols. "If your mother wasn't too pleased at your learning to ride a horse and skate and sail a boat, what will she say to this? But, Ruth, flying is certainly a great adventure, and I wanted you to have the experience—since I knew you wished it."

She looked delightedly at her companion. Master of Hounds, skilled in many forms of athletics, this father of hers was certainly a sport in every sense of the word. He had helped make the outdoor world hers and had taught her the fine points of many different sports. Impulsively she confided, "Dad, I'm going to learn to fly. I've got to because I was scared up there."

Not that evening nor for a long time to come did she mention this ambition to anyone else. Flung against the decorous background of the household at Rye, New York, it seemed sheer madness. Besides, she was engaged in a more immediate conflict with her mother's conventional

ideas. Just graduated from the Masters' School at Dobbs
Ferry, Ruth was supposed now to employ the cultural
graces she had acquired for the benefit of home and
family. Her wish to go to college met firm material opposition, but in secret the girl had been preparing for her entrance examinations. A first failure to meet the full standard
only made her the more determined to overcome the educational handicap due to many changes of secondary
schools. To Wellesley College she meant to go. And she
did!

In the end Mrs. Nichols had to accept the inevitable.
She even learned to share her husband's pride in their
daughter's success in college reports, in dramatics, and in
campus activities. Ruth's choice of studies was characteristic
—one course in each of the sciences, but major courses in
Bible and sociology. What delighted her father most was
the championship she won in riding. When, at the end of
her junior year, Mr. Nichols fell seriously ill, she decided to
stay at home, with a silent proviso of her own that she
would return to finish the course.

That intermediate twelve months was not spent as a
parlor ornament. Ruth took courses in domestic science,
studied music, and served as a Junior League volunteer at
the New York Tonsil Hospital. She dipped into the whirl
of debutante life in New York and went to parties in Rye.
But never could she forget the challenge tossed at her high
in the air, and when she joined her family in Florida for

a winter season at Palm Beach she stepped out to meet it. This beginning was followed by other trials in the summer both at Long Island Sound and at Lake George, New York, where the Nichols family spent many weeks.

All in the utmost secrecy, of course! She would be gone for hours and would stroll back with an air of utter unconcern. In a household where two lively boys, a little tomboy girl, and an ailing man were taking the attention of Mrs. Nichols, the girl's absences were ignored. Only her hands threatened betrayal. They belonged not to the lady she was supposed to be, but to a hard-working mechanic. Broken and blackened fingernails and hardening palms revealed some strange activity.

Every day down by the water the college junior, clad in overalls, was climbing into a seaplane, discovering its ways, handling its mechanism. In short, she was learning to fly, and "Captain" Harry Rogers, as he was affectionately called, a famous old flying-boat pilot, was her teacher. Admiring her cool head, her keen eyes, her alert intelligence and, above all, her determination, the pilot gave her every encouragement.

Early in the fall, when the family returned to Rye, came the culmination of these odd bits of training. "Captain" Rogers allowed her to use one of his Curtiss type Seagull models anchored in the harbor of Greenwich, Connecticut. Early one morning she took her first solo flight.

It was an unforgettable experience. To be the only one in the craft with the skill to guide it—this was a heady draught of freedom and power. Now at last she had met the challenge. She had not been afraid, but instead had felt only steady certainty and firm control.

This conquest was too glorious for secrecy. When Ruth returned to Rye that morning it was still early and she found the family at breakfast. For a second she paused in the doorway of the dining room with an aura of triumph about her. "Dad! Mother!" she cried. "I've taken up a plane all by myself. I can fly!"

Logically a first solo flight is followed by more and more until the flier is officially qualified. But Ruth Nichols was not one to leave a task unfinished, and she had to take that degree at Wellesley. After graduation her parents felt that a round-the-world trip was in order. But before she started on that momentous journey she took her flying test before the American representative of the Fédération Aéronautique Internationale. This was given at Port Washington Harbor in the summer of 1924.

"Captain" Rogers was on hand to give her last directions. It was a stiff test even for a far more experienced flier. At an altitude of 8,000 feet the plane had to be flown a specified triangular course for an hour. Then on the descent at 5,000 feet the pilot was not merely to pull back the throttle as usual, but actually to cut off the motor. Landing had to be made within 300 feet of a certain picked mark. In this case it was a millionaire's yacht anchored in the harbor.

"Flying the course was not so different from other flights I'd taken," Ruth said afterward, "except that it was awfully cold at that height in an open cockpit plane. But when I started to come down I couldn't find the yacht. There were many boats in the harbor and they all looked like rows of toys from 5,000 feet in the air. I could see no glistening white yacht with two yellow funnels, but as I came down and could see the water spaces between the boats, I thought I could identify the place where the yacht was anchored, and selected a marker which was a mere speck on the water. Down I came as steadily as possible, struck the water at fairly high speed, but lost momentum not far from my chosen speck. The speck moved. It was a canoe, and in it were the F.A.I. inspector and my proud instructor. When the 'Captain' congratulated me on picking out the canoe, I discovered that the yacht had been moved during my hour in the air and that the canoe had taken its place. I never had a better stroke of luck than that."

Just to keep her from feeling too much like a goddess, however, fate decreed a bit of ill-fortune for the second part of the test. At 300 feet the girl had to fly a figure eight ten times. The maneuver went so swiftly that she lost count and came down after the ninth figure. The meticulous official would not accept this shortage. Although it was getting dusk, he declared the flier must go up again and make ten successive loops.

"Captain" Rogers growled with blasphemous anger: "You know she can make ten eights if she's made nine. I'm not wasting my time watching any more. You can stay for such nonsense if you like, but I'm leaving."

Ruth, however, made no protest. Up she went, and this time kept accurate count. The F.A.I. certificate she won then makes her today the earliest licensed woman flier who is still actively piloting a plane.

That success made a perfect overture for her round-the-world trip. In many countries of the East the traveler met hair-raising adventures, but only in Honolulu and in France was she able to find planes in which to fly. As copilot she flew a big Army Martin bomber in Hawaii, and even more precious experiences awaited her in France. At the famous field, Le Bourget, she was allowed to pilot a large transport plane across the English Channel—a demonstration reported with journalistic fanfare in the Paris papers.

Thus the "old Wellesley grad" of one year returned home with flying points added to her score. Deciding to take up land piloting, she practiced for some time and then made her official solo flight from Roosevelt Field. Events of the year 1926, however, made regular flying impossible. The failing health of Mr. Nichols compelled him to retire, and in the conviction that she must go into business, his daughter Ruth took a position in the National City Bank as assistant to Mrs. Natalie Laimbeer, the first American woman to become bank treasurer. It was excellent training. But as the months went by Ruth thought the future looked prosaic and unrewarding except to a girl who not only had powerful contacts, but would use them to make sales.

"I just can't do that!" she thought. Torn between the military, adventurous, and sporting heritage from her father and her maternal Quaker ancestry with its seriousness of purpose, Ruth pondered on her own possibilities.

In the late summer the youthful pilot went abroad as companion to a young girl who was suffering from a nervous breakdown. Whenever possible the two traveled by plane and were not even discouraged by a crash en route to Vienna. In spite of misgivings about the effect of the accident upon her young charge, Ruth found that flying proved helpful to disordered nerves. In England the two hired a plane and, with bags tied on the wings, scooted from Devonshire to Scotland like a very modern pair of gypsies.

What Ruth wanted was activity which combined love of adventure with a conscientious sense of obligation. She was now convinced that aviation had a limitless future and that American women must become more air-minded. She wanted to help make them so and thought she might earn more money to assist her family as a flier than as either a social worker or a businesswoman. Neither a mere job nor money for its own sake had the slightest appeal to her.

Her first step was to obtain a position with the Fairchild Aviation Company. In their sales and promotion department she was the first woman executive for a million-dollar corporation. As such she had her first experience in making public addresses. In the summer of 1927 her excellence as a pilot received authoritative recognition. At last the U.S. Government had taken over the responsibility of issuing licenses under the Department of Commerce. Ruth chanced to be a little later than Phoebe Omlie in making application for a test and was therefore the second American woman to be licensed. She was, however, the first seaplane pilot and held an airplane and engine mechanic's certificate.

With Lindbergh's flight in May of that year America seemed to wake up fully to excited interest in aviation. When she met Amelia Earhart, Ruth Nichols discussed with her plans for a woman's flying organization. But the time didn't yet seem ripe for it. Amelia's own fame had still to be won. It was Ruth Nichols who leaped into the front rank of women fliers by sharing an expedition of prime importance undertaken for the Fairchild Corporation.

The flight began at dawn on January 5, 1928, from Rockaway Naval Station. Ruth was one of a trio consisting of Major N. K. Lee and "Captain" Harry Rogers, her old friend and teacher, who was now president of the Rogers Airlines. The test of the Fairchild seaplane was the first nonstop flight from New York to Miami, Florida.

"Ours was a single cockpit plane," Miss Nichols says, "and that meant that whenever Harry changed places with me, he had to jump quickly over the back of the seat without letting go of the stick until I had slid into place from the other side and had taken hold of it. It was the same when he took my place."

It was a clear, chilly morning with a north wind blowing. In an almost cloudless sky the rising sun cast rainbow reflections on the water. Gradually the sky changed to deep blue. The seaplane traveled at an altitude of 3,500 feet and, in order to conserve gasoline, kept up an average speed of about 95 miles an hour. As the fliers passed the Virginia coast they watched ice and snow gradually disappear. From Cape Lookout they flew in an almost direct line to Jacksonville and then along the seacoast to Miami; 12 hours and 15 minutes was the flying time. Today that time has been almost cut in two, but sixteen years ago the achievement was brilliant.

"Only a poet could describe the sunset over Daytona Beach," Ruth wrote in her newspaper account of the flight. "Afterward, when only a golden glow was left in the west, we were on the alert for the smudge fires which were to be lighted to guide us after dark. As soon as we sighted them we were sure of our course into Miami. At first the town seemed only a blaze of lights and we were nearly over the Causeway before I recognized it."

To the wildly enthusiastic crowds which greeted the party Miss Nichols said, "This was just an ordinary trip— not remarkable at all." But the whole country thought otherwise and gave the fliers as much publicity as if they had crossed the Atlantic. They deserved it, because they had really blazed a trail for the coastal air line which was opened up a few years later.

Sales of private planes in these boom years were mounting high. To inspire serious amateur flying, Ruth Nichols was asked to assist in a new effort. Plans were afoot to establish social flying clubs of a sort which in England had already dotted the landscape with fine landing fields and well-equipped clubhouses and grounds. Ruth became one of the organizers of aviation country clubs and through her business and social connections was able to help both in raising funds and in forming interested groups.

Given a large luncheon in Washington by Government officials, aviators, and socially distinguished men, she was sent on a nation-wide tour to promote the plan. Several exclusive and delightful aviation clubs stand as tangible results of this effort. As for Ruth's 12,000 miles of solo touring in her Curtiss Fledgling, accompanied by an escorting plane, it made her famous in every part of the land. She was the first woman to land in each of the forty-eight states. Indeed, only her escort pilot, Colonel Lindbergh, and Admiral Byrd had ever made such a record.

During the latter part of the tour she entered the first women's transcontinental derby. The "Powder-puff" Derby of 1929 proved unlucky for the flier. She was in third place in the next-to-last lap. Then the combination of a cross wind and the presence in the runway of a tractor left there by some fearful carelessness brought her plane crashing down and put her out of the race. However, other competitive events followed that year and the next in which Ruth generally took first or second place.

Whatever the famous derby failed to do, it accomplished two important results. First, it brought together the best and most experienced pilots and the most gifted novices. Second, it marked the serious entry of women into the realm of racing. Now the plan for a flying club which Amelia Earhart and Miss Nichols had shelved was put forward by others. Marjorie Brown, a licensed pilot, brought the idea before the girls in the Curtiss-Wright Flying Service. That Curtiss women's department proceeded to put the wheels in motion. Neva Paris, Opal Kunz, Betty Gillies, and the attractive stunt flier from Texas, Frances Harrell Marsalis, combined to organize a first meeting of women pilots at Valley Stream, Long Island. Because there were available ninety-nine women fliers for charter membership, the club was called the Ninety Nines.

From behind the scenes Amelia Earhart and Ruth Nichols helped shape the policy, constitution, and progress of the new organization. A nonflier named Clara Studer carried on most of the correspondence and clerical work and edited the news letter issued by the Ninety Nines for many years. Ruth was chairman of the first constitution committee. Louise Thaden, winner of the first women's derby, was elected the first national chairman and later Amelia Earhart was the first national president. This club has always been the official group among women fliers. Today it has some 500 members from all over the world, but it still retains its original name.

National prestige and leadership among women fliers would have been rewarding to Ruth Nichols in full measure except for one thing. She had to have money for her family as well as for flying. From the enormous enthusiasm aroused by Amelia Earhart's transatlantic flight as a passenger in

1928 it was plain that a solo flight meant a short cut to financial security. Ruth made up her mind to be the first woman ever to pilot a plane solo across the ocean. That would bring fortune. As the plan matured in her mind, she realized that to carry it out she needed both money and experience. Late in 1930 she convinced Powel Crosley, president of the Crosley Radio Corporation, that he should lend her his airplane to set three world records for women. That these were merely first steps in a more daring adventure was known only to her technical adviser, Colonel Clarence Chamberlin, and a few coworkers.

After long practice in the powerful plane, the flier was ready for her first attempt. This was to establish a new cross-country record for women. On November 24, 1930, butting severe head winds and storms most of the way, she flew from East coast to West coast in 16 hours, 59½ minutes' flying time. By eight hours she beat the former mark of Mrs. J. M. Keith-Miller, Australian aviatrix. Yet many troubles had grounded the plane so often that the trip required a week. Disappointed with so many delays on the way West, Ruth determined to make a one-stop return flight. On December 9, before reporters could get to the field, she was off. With only an overnight stop she made the West-to-East flight in 13 hours, 21 minutes. As the first woman to make a one-stop flight across the continent she had bettered the flying time of Charles Lindbergh the previous year and came within an hour of Frank Hawks's great record. Such success was the more remarkable because snowstorms and gales over the Rockies drove the girl up 20,000 feet, and rain squalls east of St. Louis beat on the windshield with such fury that she often had to peer straight down through the side window of the cabin in order to judge her height above the ground.

Ruth was jubilant when she put her plane down at Roosevelt Field that memorable afternoon. And in all the crowd nobody was so proud as Mrs. Nichols, long completely reconciled to the career of her famous daughter.

Now came the second test. At that time a flight across the ocean required maximum height, maximum load, and maximum speed. Colonel Chamberlin advised that the plane be tried out for every requirement. Completely redesigned at Colonel Chamberlin's Jersey City factory, it was made ready for a new altitude record for women. On March 6, 1931, the flier waved good-by to her brother and friends and zoomed off from the Jersey City airport. She finally reached 28,743 feet, a mark far higher than had been set by any woman flier.

Incredible was the cold in that region of thin air. As it registered 60 degrees below zero the thermometer broke. Later, when obliged to remove the oxygen tube from her mouth in order to reach back and turn on another gas tank, Ruth's tongue froze. Properly equipped modern planes offer fliers no such torture and risk, and Ruth's record is the standard altitude for stratoliners.

Up to that time a heavy plane with huge wingspread was used for endurance cross-country flights, a short-winged type for speed, and a light plane with powerful motor for high altitude. Miss Nichols and Chamberlin believed that the Lockheed Vega with its 650-horsepower Wasp engine would serve for all three purposes. Having proved the plane could set cross-country and altitude records, Ruth now put it through its speed trial.

On April 13, 1931, at six in the morning, she took off from Grosse Isle Airport, Detroit. She made two round trips over the course with an average timing of 210.704 miles. By thus shattering Amelia Earhart's record Ruth set a new one for women pilots. It was a great triumph for the Lockheed, and the changes in its design which Ruth had helped Chamberlin make had a permanent influence on other models. Wiley Post admitted he used the basic ideas for his famous around-the-world plane.

Now on the basis of these records, thoroughly experienced in every phase of supreme performance, Ruth was ready for the great flight. As copilots several women had tried to fly across the ocean—Mrs. Francis Grayson, Countess Lowenstein-Wertheim, Hon. Elsie Mackay, and Mrs. Bore Harte—but all had been lost. Only luck made possible the rescue of Ruth Elder, who had bravely tried the feat with her pilot. One can imagine, therefore, the stir created when Miss Nichols finally announced her plan. The first woman to attempt a solo flight across the ocean! Immediately contracts amounting to many thousands of dollars were offered her by various firms and organizations —provided she was successful. At that moment several men

RUTH NICHOLS with the Lockheed Vega in which she became the only woman in the world to have set three maximum international records for women in altitude, speed and long distance.

A crowd greets her on arrival with her famous plane.

distinguished in aviation were also poised ready for flight, but interest centered on the attempt of Miss Nichols.

After nearly three weeks of final preparations and strained waiting, reports of favorable weather reached Floyd Bennett Field. Action was instantaneous. Ruth's plane was wheeled out and Clarence Chamberlin climbed in to warm up the supercharged motor. The plane had one of the first controllable pitch propellers ever built, which is now permanently on exhibition at the Smithsonian Institution in Washington. It eliminated some of the danger of taking off with a heavy load by permitting raising and lowering the pitch by means of a lever in the cockpit. Now the device is standard equipment for most of the larger ships.

It was nearly three-thirty on the afternoon of June 21 when Ruth left the field. As if on a casual flight, she wore a simple knitted sports suit, and she chatted with Chamberlin in a manner utterly lighthearted. Then with a wave of the hand to the assembled crowd of fliers, mechanics, and friends, the young woman climbed into her plane and disposed herself as comfortably as possible in a cockpit crowded with thirty instruments, equipment, and energy supplies.

Down to the far end of the take-off strip she taxied. Then around to face a strong wind, with motor roaring and the propeller a gleaming disk in the sunlight. At first the ship started sluggishly. Gathering more and more speed, it left the ground and, as spectators held their breath, climbed slowly up to flash its golden wings against a dazzling blue sky. A squadron of Navy Helldivers appeared to escort the plane around the Statue of Liberty—Ruth's farewell to New York. Soon the escort turned back.

On alone for the great adventure went the plane with a gray-eyed girl at the controls. Before she reached Newfoundland she had to land for the night. Such was the stipulation of the insurance company. Deciding against descent at Portland, she went on to Saint John's, Newfoundland, which she reached at the end of the day.

Brilliant, level sunrays made vision difficult. The landing field, already edged with a crowd gathered to welcome her, was easy to spot. She made a complete circle of it, and as she did so misgiving smote her. She had been misinformed about the size of the airport. Her experienced eye told her that the runway was not long enough for the speed of her fast plane. Her landing gear, built for speed, had not the rugged strength necessary for the "pancake" landing required on so short a field. Could she make it? Believing she had lost sufficient momentum in the descent, she decided to try.

Almost down—and then she knew. Ahead at the end of the field rose up a wooded clifflike hill. The momentum was still too great. She would hit that cliff head on. It meant certain death. With but a second for decision she gave the plane "full gun" and shot upward. For a moment she was sure she could just skim over the top of the cliff. Then a projecting crag loomed above her. The crash of steel on rock was echoed by cries of horror from the on-lookers below.

With undercarriage stripped off, the plane hung danger-ously pinioned by rocks and trees. Up the hill a dozen men raced panting. First to reach the plane was a cameraman. He saw the flier dragging herself out of the cockpit to the top of the plane. "I was afraid it would take fire!" she gasped.

Two of the rescuers made a seat of their locked hands and carried the girl through woods and over rocks. Ar-rived at the field, she tried to reassure everyone. No ambu-lance for her, she said. A taxi, a hotel room, and a bit of rest were all she needed. But halfway back over the cor-duroy road an ambulance came clanging and took the flier straight to the hospital. There it was found that she had fractured four vertebrae and crushed a fifth.

On the examination table in the emergency room she held up doctors and nurses to wire her mother at Rye, "All I did was to wrench my back. All O.K. Awfully sorry about crashing, but will do it next time."

In all her twenty-two years of flying up to the present day this was the only injury Ruth suffered when piloting herself. Physical pain and nervous shock were bad enough. Devastating disappointment was almost worse, for the ex-pensive plane was smashed and she had neither the strength nor the necessary backing to make another at-tempt. Yet such was the pluck and tenacity of this flier that three months later, still wearing a plaster cast, she was out to set a women's long-distance world record. The run was 1,997 miles from San Francisco to Louisville, Kentucky. Encased in a plaster cast and steel corset, Ruth streaked victoriously into Louisville with a new 14-hour record in tow. She had flown eleven hours in darkness and eight of these at 15,000 feet. This record made Miss Nichols the only woman in the world then and today to have set all three world records for women—speed, altitude, and long distance.

At Charlotte airport she gave exhibition flying. On February 17 of the following year, 1932, she bettered the national Diesel altitude record for both men and women. In spite of intense cold she drove her "Flying Furnace" up to 19,928 feet. When, finally, by easy spirals, she floated the ship to the ground, there was only half a gallon of fuel left in the fifty-gallon tank.

To summarize the flier's accomplishments in the next few years presents an almost fantastic picture. She became first woman pilot of an air line when serving as reserve pilot for New York and New England Airways. As "Air Ambassadress" for five million women she made a 3,000-mile tour about the country to promote the 30th Women's International Congress and was assigned the United States Championship for Women by the International League of Aviators.

In the summer of '32, still keeping in mind the possibility of a transatlantic flight, she went abroad for many flights along the coasts of France, England, and Ireland. Bad weather forced her to give up her ocean flight that year, and by the time another year came around there was not sufficient public interest to support such a project. Instead, Ruth laid tentative plans for an air voyage from Honolulu to San Francisco. She found, however, that, working alone as she must, it was too difficult to raise the money for the flight. When Amelia Earhart, through the efforts of her clever promoter-husband, succeeded in undertaking the same flight, Ruth gave her friend some of her maps in order to make the chances more certain.

Ruth's own career now took a new turn. She embarked upon an educational campaign. In the fall of 1932 she lectured at a number of colleges on aerodynamics, avigation, and aerology. A tribute to her great success in this field was the honorary degree as Doctor of Sciences given her by Beaver College.

Now and then she took part in races and meets. With another pilot long skilled in twin-motor airplanes, she went on a barnstorming tour of New England to spread interest in flying. They sold short trips to people who had never been in a plane before. A serious accident cut short this happy expedition. The fliers were nearing Troy, New York, one day when apparently the pilot made a fatal mistake. The plane crashed and burst into flames. The pilot was killed and Ruth Nichols was so terribly injured that she spent seven months in bed.

When the dozen fractures she had suffered were nearly mended and Ruth was at home again, she thought only of how soon she could fly. She walked out-of-doors a little, was taken for a drive another day, and then she demanded to be taken to the airport. Nobody could stop her. In splints and bandages, she was, as she put it, "shoveled into a plane" —just to get the feel of it again. Long before the healing was complete, she was at the controls happily zooming about the skies.

To someone who called her a heroine for this comeback Ruth said, "Oh no. When a person loves flying and understands it and knows the reason for an accident, there is no fear. I could hardly wait to be in the air again. That's just natural."

Ruth Nichols never stops blazing trails. She was the first woman staff member of an aviation magazine and one of the few Americans admitted to membership in the Women's Engineering Society of Great Britain. In 1940 she was cited by the General Federation of Women's Clubs as one of the three most outstanding women air pioneers. The other two were Anne Lindbergh and that inspired flier and teacher from days of World War I, Katherine Stinson.

Personal honors always counted less for Ruth, however, than opportunities to work for happier human relations. As rumblings of World War II began to be heard, she became obsessed by one dominating idea. Aviation must be associated with constructive humanitarian values and not exclusively with the hideous destruction of military aviation or the pure materialism of commercial flying. As a

Quaker and sociologist, she was active in the Emergency Peace Campaign and used aviation as a springboard for discussing the deep causes of world unrest. Then, with closer approach of war to this country, she initiated an organization which promises to be of supreme importance.

Relief Wings is a humanitarian air service. Its activity is fourfold: The enrollment and training of surgeons and nurses for flight work, the providing of aero-medical supplies such as special-type stretchers and aero-medical kits, research and study in the care of air-borne patients, and the development of ambulance planes both by building specially equipped twin-motored planes and the conversion of private planes for air ambulance work. A corps of operators with portable radios was also part of this plan. Since war was declared by Germany and Japan, however, this service has been postponed "for the duration." At first planes and pilots were registered directly by Relief Wings for disaster service. Now the organization works with the Civil Air Patrol.

Why was such an organization a vital idea? The reason lies in the necessity of bringing aid swiftly to individuals or to cities and towns afflicted by disasters, whether in peace or in war. During the hurricane of 1938 many small towns in New England were entirely cut off from the outside world. Had private planes been ready with nurses, doctors, and radio operators, registered and accustomed to flying, to speed to marooned towns where there were no airports or with none large enough to accommodate large air liners, help, supplies, and information as to added needs could have been rushed to marooned, suffering communities. Natural catastrophes always threaten, and some aviation agency must be ready to serve the sufferers. During wartime this necessity becomes imperative. Should the many isolated airplane factories be bombed, it would take a fleet of ambulance planes to handle human needs.

It was Ruth Nichols who had the imagination and the human sympathy to visualize such a service. This is the only large country in the world which has no twin-motored ambulance planes. Moreover, despite the fact that we have manufactured just such planes for other countries, that we have sent millions to aid Allied nations, and that we have spent fabulous sums here, the United States Army and Navy together could boast, at our entrance into the war, of only a few single-motored, specially equipped air ambulances. Conferences with the Red Cross have brought out the fact that this agency has not seen fit to own planes of its own or to co-ordinate existing civilian aviation facilities.

Relief Wings is gathering support which comes to any unit answering a vital need. People are beginning to realize how many individuals have needed an air ambulance, even in peacetime. Some 3,000 patients have been flown to and from the Mayo Clinic at various times. Some of these patients not only could have been made more comfortable with air-ambulance equipment, but others were actually protected from further injury due to lack of suitable equipment. From the research already accomplished in the treatment of air-borne patients can be shown how much has to be learned. Here are several cases. Anyone who has a concussion of the head must be given oxygen at all levels in a plane. Otherwise, dire results occur. Those with penetrating wounds of the abdomen are exposed to great risk due to expansion of gases as a result of change in altitude and pressure and must be treated accordingly. Some patients should not be flown at all, such as those with infected sinuses. Miss Nichols herself knows what it is to be the victim of an accident, flown from the scene to a hospital, and how important it is to have a nurse who does not become emotionally disturbed by flight.

Sponsored by leaders of all large aviation groups in the United States, Relief Wings has sections with jurisdiction over thirty-eight states. Money is being raised for the air-ambulance service as well as disaster relief by air. It bids fair to make a great contribution to modern aviation.

Such a cause offers Ruth Nichols just that combination of adventure and idealism she had long been trying to weld into her career. Pioneering to serve humanity and alleviate suffering through aviation—there is complete fulfillment for a great-hearted woman. The ultimate goal of this distinguished flier is not winning laurels, but giving service.

Louise Thaden

THE ARKANSAS TRAVELER GOES HOME

☆

Nᴇᴀʀʟʏ six o'clock of September 4, 1936. Here at Mines Field, near Los Angeles, officials' eyes were straining toward the east. Which of the famous pilots who had started from New York would appear first? The question was vital from the standpoint of both money and prestige. For this great Bendix Race, first of the National Air Races, offered $4,500 to the first contestant crossing the finish line and $10,500 to the four closest rivals.

"I'm betting on Benny Howard!" That's what was heard most frequently. And why not? Hadn't Howard won both Bendix and Thompson races the previous year? Wasn't his Mr. Mulligan a superb plane stepped up for a high rating? However, a number of enthusiasts were staking their faith on Joe Jacobson. "Wonderful flier with a special Northrop noted for speed!" Other contestants also had their supporters. All agreed it was a shame that one of the best fliers had cracked up his Wedell-Williams racer even before the start.

Three women had entered the Bendix Race. They were Amelia Earhart in her Lockheed Electra, Laura Ingalls in a Lockheed Orion, and Louise Thaden in a Beechcraft. But did any of the experts at Mines Field speak of them? No indeed. Of course it was a fine gesture to let women compete this year. Sort of chivalrous; a tip of the hat to a trio of very good pilots! But as for one of them winning the Bendix Race from New York to the California coast— well, hardly!

While the experts wrangled and fidgeted and listened for the telephone to buzz, the crowd was living in the excitement of the moment. For its benefit a flying circus had been set going—loops and parachute jumps and air stunts of every description. But suddenly from the field manager's office bad news began to circulate. Benny Howard and his wife Maxine had cracked up in New Mexico and were badly hurt. That horror was followed by a report that Joe Jacobson, caught in a terrific thunderstorm over Kansas, had had to bail out. Officials and the relatives and friends of the racers were in a state of tension. With conditions such as to knock out the favorites, news of the others was awaited with dread.

Suddenly in the eastern sky a blue-and-white plane appeared. By the crowd, absorbed in watching a parachute jump, the arrival went unobserved. But the officials rushed forward. They watched the plane zoom over the finish line. Eyes stared. Lips parted in amazement. Could it be? A woman winning the Bendix Trophy? Then, as the plane taxied to a stop, a shout rang over the field: "It's Louise Thaden! Louise Thaden and Blanche Noyes are winners!"

The crowd rushed pell-mell about the plane. Out of its tiny cabin stuffed with extra gas tanks two young women squeezed themselves with difficulty. First to emerge was Blanche Wilcox Noyes, an appealing slip of a person whose smile smoothed out the tired lines from her charming face. Next came—was it one of the Valkyries? Tall, statuesque, with fine head molded by the thick, short brown hair fitting it like a casque, Louise Thaden had the very air of riding the clouds to victory.

But she couldn't believe what those cheers meant. She and Blanche faced the tumult with blank faces. "You mean we've won—we?" drawled Louise. "Why—it can't be true!"

For an instant she was too amazed for joy. Then, with dark head flung back, she became a figure of exultation. In a soft, unhurried tone she told the story to the reporters. Even before she and Blanche had reached Cleveland, their radio had stopped working. From that point on bad weather and severe head winds beset them hour by hour. They made the best speed possible, but supposed, of course, they would be at the tail end of the race.

Mrs. Thaden's time was 14 hours, 49 minutes. She was almost an hour ahead of Laura Ingalls, second to arrive. Third was William Gulick; fourth, George Pomeroy; and fifth, Amelia Earhart with Helen Richey. The last two had been delayed by a long battle with a flapping emergency door.

After being photographed and interviewed, the winners hurried away. Louise had to call long-distance at once and tell the news to her husband, Herbert Thaden. Even as she heard his excited tones she could hardly believe what she was saying. "Yes, both prizes, honey! Blanche and I divide $4,500 from the Bendix purse and $2,500 more as the women's prize. Isn't that swell, Herby?"

Walking back to the field to wait for the other racers, the young woman pictured her husband's elation. Then her thoughts flew from Maryland to her mother's home in Arkansas where her two children were staying: An excited little boy jumping around his grandmother with a dozen questions about the race; even the three-year-old would catch the ferment and clap her fat little hands. Some day, with luck, thought Louise, the Thaden family would be reunited. Today's prize was a good omen. Then she wondered whether she could ever give up aviation.

That was a love antedating all others. It began when she was Louise McPhetridge, a girl from a small Arkansas town, at work one summer in Wichita, Kansas. The city was headquarters for a large airplane factory called the Travelair Company, of which Walter Beech was president. Beech was producing three place biplanes for the growing aviation industry and their test flights added constant stimulus to the interest in flying so characteristic of an airport city.

One day the college sophomore from Arkansas went out

to the flying field. Instantly she was seized by an irresistible fascination. Gone were the dreams of other vocations— journalism, medicine, business. From then on she lived for one purpose only, to learn to fly.

After another year at the university the girl returned to her job in Wichita. Her employer, who was interested in the Travelair Company, knew Walter Beech very well. When he learned the ambition cherished by Louise Mc-Phetridge, he set out to help her. It wasn't long before Beech gave the girl a job as assistant to his West coast distributor at San Francisco. Off to the Golden Gate she went as passenger in a Travelair plane.

The work she had to do was taxing; but each long day of typing and keeping records ended in glory; for then came the flying lesson. She put so much zest into learning that she won a private pilot's license in a very short time. By the end of 1927, therefore, her feet were on the ladder leading to the sky.

To climb the second rung required a transport license. That demanded a credit of two hundred hours' piloting— no easy matter for a busy wage earner. Only backing from her company made it possible. Aside from instruction and special coaching, the girl was given the use of a Travelair Hispano-Suiza 180-horsepower plane. In it on December 7, 1928, she rose from the field at Oakland to an altitude of 20,200 feet. Since at that time the woman's record was held by Lady Heath, the famous British aviatrix, at 16,438 feet, this was a smashing victory. Three months later, on March 16, 1929, Louise made a new endurance record by staying in the air 22 consecutive hours, 3 minutes, 28 seconds.

Naturally these records had brought prestige both to her and to the Travelair Company. Its president now urged her to try for a speed record and provided her with a plane equipped with special racing wings. By reaching a speed of 156 miles an hour, the youthful flier again proved herself a champion. When she successfully passed her test for the transport license, she became the fourth woman transport pilot in the United States.

"And now what wreath of victory are you going to snatch, Louise?"

Half teasing, half anxious was the tone of this question. The young man who asked it had often watched the girl fly, had sometimes flown with her, and for many months had talked with her about aviation whenever he had a chance. He was Herbert Thaden, an ex-Army pilot and engineer, who was building an all-metal plane in a hangar at San Francisco. Often over lobsters down at Fisherman's Wharf the two young fliers had discussed their careers. But this time it was different. For Louise, equipped and ready, stood on the threshold of new achievement.

"I think," she answered her companion, "that I'll go home for a while and see my family. Then I'll decide what is the best thing to do."

At this announcement Herbert Thaden's face grew stormy with protest. The one congenial girl he knew, who shared his devotion to flying—no, he couldn't let her leave San Francisco! Suddenly he knew that between them there must be no thought of parting. What did lack of money matter? What did all their talk about separate ambitions amount to? Something else counted more. The young man's ardor sent common sense into a tail spin. Deciding to set a speed record all their own, the pair set off for Nevada. At Reno, where so many marriages end, the marriage of Louise and Herbert Thaden began.

Their happiness was given a filip that summer by another success for Louise. Among twenty of the best women aviators she entered the Women's Air Derby from Santa Monica, California, to Cleveland, where the National Air Races were to be held. Mrs. Thaden had persuaded Walter Beech to build a special Travelair plane for the event. When, however, she set it down on the Santa Monica field for the start, she realized that her ship was by no means the fastest of the lot.

Gathered at this scenic spot were women pilots from every part of the country. Some like the Australian, Mrs. Keith-Miller, Amelia Earhart, and Ruth Nichols were already famous. Others—Blanche Noyes and Gladys O'Donnell, for example—were just entering aviation. This derby proved to be a race historic for its strange mixture of comic and tragic elements. One of the best-loved women pilots, Marvel Crosson, an excellent flier herself and sister of the famous Joe Crosson, head of aviation in Alaska, was doomed to meet her death in the course of the ill-fated race.

Then and there Louise Thaden formed lasting friendships, particularly with Amelia Earhart, Blanche Noyes, and Ruth Nichols. She was interested, also, in meeting for the first time Ruth Elder. That pretty young woman had practically lived in the limelight for the two previous years. After a brief career in flying she had set out, on October 12, 1927, to cross the Atlantic as copilot with George Haldeman. A few hundred miles from the European coast the plane was forced down, but the fliers were rescued. Wildly welcomed in Paris and in New York on her return, Miss Elder had won an international reputation for supreme courage, and for a long time she remained a feature of the news. At the Women's Derby the bright red plane she flew sustained her reputation for drama. It also gave her a bad moment. In one lap of the race she was forced to land. The field she had to use was filled with horned cattle in the midst of which she set down her scarlet ship. When Miss Elder was asked what she did about that, she replied, "I prayed. I said, 'Dear Lord, let them all be cows.'"

That incident was only one of many mishaps which dogged the women pilots between California and Cleveland. Louise Thaden, however, sailed steadily on her way and crossed the line first of all the contestants. Her elapsed time was 21 hours, 19 minutes, 2 seconds. After that triumph the young woman became a marked figure in American aviation.

The derby was run on August 27, 1929. Shortly after that Herbert Thaden was transferred to an airplane factory in

Pittsburgh, and for the first three years of residence in the smoky city his wife disappeared from the air meets. She was, however, elected national chairman of the newly organized Ninety Nines. Moreover, she was constantly flying to demonstrate the planes her husband built. Louise did take a little time off to soar into the realm of motherhood and soon she and Herbert went cruising in the air with a very young passenger by the name of Bill Thaden. He became as familiar with a cockpit as most babies are with a nursery pen.

In 1932 a firm in Baltimore took over the development of metal ships. Of course that meant another shift for the Thadens to the Maryland city. But in August of that same year Louise once more entered the aviation lists. A plan was worked out in combination with another charter member of the Ninety Nines named Mrs. Frances Harrell Marsalis. Frances worked for the Curtiss Exhibition Company and competed in most of the women's aviation meets. Her stunts had become famous, but she was ambitious to establish new records. With this experienced and dashing flier Louise Thaden was invited to combine in a record-breaking refueling endurance flight.

The mark they had to beat was 123 hours in the air. It had been set by two clever pilots, Miss Bobbie Trout and Miss Edna Mae Cooper. Such a flight means that no descent is made to refuel. Gasoline is transferred to the plane's tank by another plane which flies above it. This is always a tricky maneuver and fraught with danger. The real value of an endurance test is to establish unquestionable proof of the plane's quality.

William Marsalis, pilot and instructor, was with his wife on Curtiss Field to greet the flier from Baltimore. The project, however, was under the supervision of Viola Gentry. It was she who engaged the pilots for delivering fuel and supplies and secured regular weather reports through the airport manager. One of Miss Gentry's inspirations was to improvise a bed on the back seat of the plane by means of an air mattress.

"You realize, of course," she said to the contestants, "that you'll have to stay up in the sky at least a week to break the record."

Why, of course they were prepared for that. So they had replied. But when, after every final check was made and they had sailed off, the two pilots faced their first all-night session, they began to realize what was in store for them.

All the more disappointing was the accident that befell them next morning. Their contact man, flying above them, was lowering a bucket containing food and drink. Somehow it struck a wing and tore a bad hole in the fabric. There was nothing for it but to come down, have the wing repaired, and start all over again. Only a refreshing all-night sleep made up to them for that bitter loss of eighteen hours.

It was on a Sunday morning when once more they rose into the air. For several hours bumpy weather offered a discouraging start, but by afternoon the air was somewhat calmer. That evening they went cruising over Floyd Bennett Field to look down upon the many sailboats at anchor on the glistening water. Upon their return to take on fuel they found the air so rough that only after three attempts could they get close enough to the refueling plane to establish contact.

An hour later, however, the moon came up. "Isn't it really marvelous," said Louise happily, "to be up here all night long? Look how the lights down there have changed since the moon's up. They're just tiny sparks."

Even more magnificent is dawn watched from the sky. Then the whole world can be seen flushing and turning gold. Many such moments of beauty were shared every day. But the keenest response to them was not proof against exhaustion. Hour after hour, morning, afternoon, and night, the motor droned on. Food dropped from the plane above never tasted like a real meal. Limbs were cramped. Eyes smarted. Backs ached. The aimless movement created an almost unbearable monotony.

On the fifth day, however, almost too much excitement occurred. First, they had to outride a fierce storm. Then about noon, when Louise was at the controls, both girls cried out in one anguished voice, "Look at the oil gauge!" The needle was dropping rapidly. Frances pumped with all her might, but without effect. The gauge showed an empty tank. Wildly the two shouted commands at each other. Frances scrambled into the rear of the plane to search. At last she discovered the trouble. The drain valve was open. Evidently one of them had knocked it open when moving luggage.

"How much oil in the cans?" cried Louise.

There was just enough to last until the next refueling. The valve was closed. All the cans were emptied into the tank. Slowly the needle began to rise once more. Sinking back into their seats, the frightened girls gasped out, "Saved again!"

On the seventh day a thrilling sight met their eyes. It was the passage of a plane escorted across the sky like royalty. James A. Mollinson, who had just flown across the Atlantic, was arriving in New York. What with saluting the hero and discussing his exploit, the weary fliers kept awake for some time without effort. Then the sleepiness they had learned to dread stole over them. Neither of them dared take more than a cat nap of an hour or so for fear the other pilot would go to sleep at the controls and wreck the ship. They used every imaginable device to keep each other awake through the eighth day.

At last a plane hovered over them long enough to drop a note. It was from the manager of the airport. Tearing it open, Louise shouted the news. "It's over! We've done it! We broke the record by seventy-three hours!"

The pair needed no second bidding to come down. Louise said afterward that she had never felt anything so welcome as the sensation of the wheels touching earth at last. For a few moments relief and excitement gave the pilots

Louise Thaden, first woman to win the great Bendix Air Race.

Greeting the crowds at Cleveland just after her first victory in the Women's Air Derby in 1929.

strength enough to greet Viola Gentry, pose for photographers, and say a word to the hovering reporters. Then they reeled off, to sleep the clock around.

They woke to fame. Messages were pouring in from every part of the country. The morning papers carried stories of the triumph. Lecture bureaus offered engagements. Moreover, a plan was on foot to present the victors to President Hoover in Washington. Louise Thaden, however, had just one thing in mind. Tucking her hand in Viola Gentry's arm, she said, "Help me arrange to hire a plane, there's a dear. I'm going to fly to Baltimore and see Herb and little Bill before I do anything else."

Then she and Frances Marsalis, who had shared eight days of beauty, danger, and anguishing discomfort in such isolated intimacy as no earth-bound friends ever know, went their separate ways. Not long afterward they joined each other in Washington where a gracious welcome was given them, and in the next two years they met many times at flying fields. No one was more heartsick than Mrs. Thaden when her flight companion had the tragic crash which ended a bright career.

In 1933 another baby made a foursome of the Thaden family. Patsy certainly chose her parents with reckless disregard of consequences. Before she was more than a year old Herbert and Louise had decided to enter the MacRobertson Air Races from England to Australia. For months everything centered on the plan. They ordered a special long-range plane from Walter Beech, a fast and handsome creation. But, although the fliers wrote and wired, pleaded and raged, the job could not be finished in time for entry in the race. Gone was the chance of winning money and world laurels. Gone was the entry fee. Even the plane, which had to be sold far below cost, represented a heavy loss.

Thaden fortunes were now at a low ebb. But skilled fliers can usually find ways to earn money. Placing the two children with her mother in Arkansas, Louise went back to work for Walter Beech. Her husband joined the T.W.A. Lines as pilot and consultant on design. For months the two could see little of each other.

As demonstrator of a new Beechcraft biplane, Louise toured the country. It was strenuous work. In addition to the usual strain experienced by the pilot-salesman when a novice first takes the controls, a woman suffers from a special handicap, for the average man is apt to think her ability is inferior to his. In order to prevent accidents without offending the egotism of a prospective buyer, Louise had to use much diplomacy in addition to strength and skill.

Phoebe Omlie persuaded Mrs. Thaden to take some time off in 1933 and help campaign for the proper air-marking of towns and cities. It was the newly organized Bureau of Air Commerce which was the logical agent for the important undertaking. Seventeen years before Ruth Law had tried to get the work started. Now there was hope of its being begun. In 1934 the Government was at last ready to initiate the service. They engaged three air-marking pilots. Logically enough one of them was Louise Thaden.

Flying skill, personality, firmness, and tact were all demanded by this job. First air surveys of the best site for lettering had to be made for each town. Local officials had to be persuaded to give their co-operation, and building owners had to yield the right to paint the huge letters on their roofs or water tanks. Then a plan of action had to be drawn up and the results surveyed.

Louise found the work fascinating. She was assigned the western part of the country and had to fly long distances. Since the Air Bureau lacked sufficient funds properly to service their planes, Louise had considerable difficulty with hers. Several times she was forced down. One day the motor failed four times and each time, of course, she had to land. On another occasion the glass blew out of the cabin roof, changed the lift of the plane, and caused it to land at ninety miles an hour instead of at its usual speed of sixty. Whoever had a soft job in the Government in those days it was not the air-marking pilot.

After a year's work Mrs. Thaden took up her own private career again. On July 12, 1936, she went out to make a new speed record. In a light Porterfield model with a 90-horse-power engine she flew over a 100-kilometer course at a rate of 109.58 miles an hour. This was a new speed level for women flying planes of that class. Her second venture was totally unexpected.

Blanche Wilcox Noyes had just joined the corps of fliers

at the bureau. In August Louise went with her friend to Texas to help her start this work for Uncle Sam. While they were there the Beechcraft Company suddenly offered Mrs. Thaden a plane for entry in the Bendix Trophy Race from New York to California. Louise persuaded Blanche to go as copilot, and with breathless speed they raced to New York for the start. To their joy they found Ruth Nichols on the field. Not yet sufficiently recovered from a terrible accident to take long flights herself, she had been appointed one of the official starters. Ruth's cry of "Good luck!" was in their ears as they winged off by the light of the morning stars.

Such was the varied experience behind the Valkyrie who arrived that late afternoon at Mines Field. At that moment of triumph Louise Thaden, most lovable of champions, had reached the pinnacle of fame. Her achievements won her the highest honor a flier can receive—the Harmon Trophy for the year's outstanding aviator, presented through its national committee by the International League of Aviators.

For a short time after the Bendix Race the victor returned to the Bureau of Air Commerce. Then the Beechcraft Company lured back its celebrated representative. For a year she demonstrated their planes from coast to coast and was a great success. Since then from time to time she has demonstrated planes for other airplane companies.

Never has Mrs. Thaden stopped flying. But after 1937 she followed the course she once outlined to Amelia Earhart. "I'm going to try to be with my children more in these next years," she said in her soft, slow voice. "It's time they knew me as a mother and not just as a flier."

To the partnership into which she had entered long ago in Nevada Louise brought rich gifts. Fame had come to her through her own courage and enterprise. Because it affected neither her natural modesty nor her sense of humor, it brought her a host of devoted friends. The public's appreciation of these qualities is largely due to Mrs. Thaden's gay little book about her experiences called *High, Wide, and Frightened*. But that journalistic account gives little idea of the great variety of ways in which the author added her bit to the progress of aviation. Certainly no story of its development would be complete without listing her accomplishments.

Lately she has been co-operating with Miss Ruth Nichols in Relief Wings. One could imagine in times of danger no pilot readier to serve or more quietly competent to do so. Stress and strain never cause that soft voice to rise nor the drawl to quicken. But Louise Thaden's casual manner is deceptive. She handles a plane with the instinctive mastery of a born flier. Whoever has seen her on a flying field taking her stately way toward her winged ship can never forget the image she suggests. The woman may have been born in Arkansas, but the flier sprang straight down from Valhalla.

Mae Haizlip and Blanche Wilcox Noyes

A TALE OF TWO FLIERS

☆

CIRCUS DAYS are over for aviation. It is now a profession offering many varied opportunities for both men and women. What is true of other occupations, however, is doubly true of these openings. A woman has to prove superability if she dares compete with men. Consequently, a background of racing and record-making serves her well. The present chapter concerns two feminine careers which are solid today because they rest on noteworthy past achievements.

During the year of 1929 both Mrs. Blanche Wilcox Noyes and Mrs. Mae Haizlip got their first mention in the annals of flying. They didn't know each other then and didn't even meet until some time after they were fresh-made veterans of the flying field. A common feature of their diverse experience was that each of them was taught to fly by her pilot husband.

At the close of the nineteen twenties American women had become decidedly air-minded. Hundreds of them were

traveling by passenger plane, and since the famous flight
of Charles Lindbergh they had followed air events as never
before. Amelia Earhart had flown across the ocean as
passenger. She had been the fourth woman, following
Phoebe Omlie, Ruth Nichols, and Elinor Smith, to get
a transport pilot's license from the U.S. Department of
Commerce. Women with the skill of Bobbie Trout and
Louise Thaden were making altitude, speed, and endurance
records. They had started to compete in cross-country
derbies, and Mrs. Omlie had tickled the imagination by
joining the 6,000-mile Reliability Tour. Therefore, there
were no more mental hurdles for any new-made woman
flier to leap. All that was required was determination.

In the case of Blanche Wilcox, a Cleveland girl, the will
to fly had to be stimulated. Extraordinarily pretty, active
on the stage and in journalism, she might never have connected with aviation but for attending a certain formal
dinner given to Charles Lindbergh in 1927. Naturally the
chief topic of interest among the guests was aviation. The
attractive red-haired young man who sat beside Blanche
spent the entire time from soup to nuts recounting the joys
and thrills of flying.

Dewey Noyes was a good pilot and a man of action.
Before the evening was over he had won a promise from his
charming companion to go up in his plane. She did so two
weeks later. But that was only a beginning of Dewey's
campaign. Within twelve months he had persuaded Blanche
Wilcox to marry him.

Of course that meant marrying aviation, too. First as passenger, then as learner, and finally as solo flier, Blanche Noyes took to the air. Once started, she went as fast as wings could carry her. In April 1929 she received her pilot's license, the first woman in Ohio to do so. In July she passed her Limited Commercial Test. In October she entered the first Women's Air Derby from Santa Monica to Cleveland. This was the so-called "Powder-puff Derby," which made its mark as an affair of accidents and confusions. Bad weather and insufficient experience accounted for the many groundings of the fliers. Comic incidents, however, were blotted out in the tragic crash that killed that much-beloved pilot, Marvel Crosson.

When the survivors reached Cleveland at last, the mutual discussion of their adventures went on for days. Nobody on the field ever forgot the appearance of Blanche Noyes, the fourth competitor to come in. Pale, weary, and disheveled, she was dragged to the microphone the moment she got out of her plane. Told she must say something to the waiting thousands, Blanche gasped, "Oh *boy!* But I'm glad to get here!"

Everyone laughed at the passionate eloquence of that relief. But when her story was told, laughter turned into murmurs of admiration. She had got off to a good start from Santa Monica, but about fifty miles west of Pecos, Texas, Blanche had a horrible sensation. She smelled smoke. In another moment wisps of smoke were curling into the cockpit. Keeping her plane on an even keel, she looked

about for the source of fire. It was in the baggage compartment just behind her shoulders. Perhaps a mechanic had
dropped a cigarette there. The danger was imminent. She
had to land. Down she came, hoping for luck in a landing
place. What lay below was a desert—just mesquite growth,
stones, and bushes covering rough, uneven ground. But
with smoke growing denser, she had no time to choose. As
the ship touched the earth, she felt a wheel give way and
heard the scratching and bumping of the fuselage. The
plane dipped sharply, and through the left wing stuck a
bristling finger of mesquite. Shutting off her motor, the
girl jumped out.

Her one thought was of the fire extinguisher. But the
heat was too great for her to wrench it out of its bracket.
Scooping up sand, she flung it on the smoldering flames
over and over again until finally they were smothered. At
last she could open the compartment, drag out the scorched
articles, and assure herself that not one spark was left. She
was trembling from head to foot.

Now for a look at the landing damage! By this time a
small rescue party had arrived. Somebody went to the nearest village for a blacksmith. Meanwhile, Blanche set about
mending the wing with a bit of linen and a surgeon's
needle. The blacksmith, who came equipped for work,
managed to weld together the landing gear. Sooner than
she had thought possible, Blanche was ready for the take-
off.

To this day she can't explain how she ever got enough

speed on that desert to lift the plane into the air. She did so nonetheless. Then off she sped for Wichita, Kansas, caught up with the laggards in the derby, passed some of them, and reached Cleveland fourth in line.

The report of her weird day which Mrs. Noyes submitted was brief and full of humor. But when Amelia Earhart described on the radio the results of the Women's Derby, she gave great praise to the skill and ingenuity shown by the young flier. A few months later, when Amelia and Ruth Nichols helped Marjorie Brown and a group of the Curtiss-Wright organization fliers to form the Ninety Nines, Blanche Noyes, as charter member, was present at the organization meeting in Valley Stream, Long Island. Friendships with Miss Earhart, Mrs. Thaden, and Miss Nichols begun that year were to mean much to Blanche both personally and professionally.

Strangely enough, it was an accident which that very same month of the derby brought Mrs. Mae Haizlip into the notice of the press. She could boast only of a broken finger, but the small misfortune made the news, because Mae was one of the forty-nine contestants in a 5,000-mile air tour of the United States and Canada.

Mae herself made light of the finger. What mattered was that she was getting real experience in flying. Slim, energetic, and humorous, Mae hailed from St. Louis where her husband was assistant to the head of the aviation department of a large petroleum company. Egged on by her small son Hayes, Mrs. Haizlip was determined to enter competitive flying. In those circles James Haizlip already had a reputation as a skillful, fast, and daring aviator.

Before Mrs. Haizlip could gratify her ambition, however, she was victim of another accident. This time it was serious. As one of the entrants for the Women's Dixie Derby in August 1930 she was flying from Atlanta to Washington. First she lost her way. Then she crashed near Greenwood, South Carolina. She was rushed to the hospital and did not recover consciousness for hours. When she did so, she eyed the doctor with comic wistfulness. "I suppose there's no Dixie Derby for me?" she murmured.

"Hush, don't try to talk!" said the doctor, and added soothingly, "You will recover, Mrs. Haizlip, in due time."

Mae grinned into the pillow. Just eight days later the patient was at the Curtiss-Reynolds Airport near Chicago all ready to fly another race. This was the first standardized closed event for women and therefore more important than the Dixie Derby. Mrs. Haizlip won. On the 25-mile course she made a speed of 121.08 miles an hour. It was on that field that she met for the first time a brilliant flier destined for a difficult fate. Although this young woman had placed only third in the race, she had already broken the records for loops and barrel rolls. Her name was Laura Ingalls.

In triumph Mae Haizlip returned to St. Louis. She had fooled the doctors, won a race, and carried off a $500 prize. James Haizlip proudly told her she was now a definitely accredited racer.

Meanwhile, our other heroine, Blanche Noyes, had won her transport license and was accumulating many hours in the air. One of these hours was reported in every American journal. When she and her husband were visiting Mr. John D. Rockefeller, Sr., at Irvington-on-Hudson, they so inspired the oil king with their enthusiasm that he asked to be taken up. The trip was arranged with Blanche as pilot. Rockefeller, who had never been in a plane before, expressed himself as delighted with the experience. But since her famous passenger never took another trip, Blanche was often teased about that last ride.

In 1931 Mrs. Noyes was hard at work demonstrating planes for the Great Lakes Aircraft Corporation. The assignment offered much varied flying experience. It was, however, the other woman in this synchronized picture who won fame that year. Hers was not the distinction accorded Ruth Nichols, who had just made her third world record, but Mae Haizlip's series of successes was nevertheless exceptional.

At St. Clair, Michigan, in June, she made an altitude record for the "Bantam" class of light planes. Her score was 18,097 feet. In August she entered the Transcontinental Handicap Derby from Santa Monica to Cleveland. Flying a Lambert Monocoupe, she followed Phoebe Omlie across the finish line and won second place.

This derby brought the fliers to the National Air Races. Intoxicated by the joy of winning an $1,800 purse, Mae jumped into all but two of the many events open to women at the meet. Slim and brown, sporting goggles, divided skirt, and boots, flashing her humorous smile, she looked the embodiment of energy.

From those competitions she emerged with an astonishing total of points. Flying many types of planes, she made the following score: first place in the 5-lap, 5-mile race for planes of 350-inch displacement; second place in the 510-inch displacement race; second place in the 650-inch displacement race; second place in the 800-inch displacement race; second place in the 1,000-cubic-inch displacement race; second place in the 1,875-cubic-inch displacement race; second place in the women's free-for-all on a 5-lap, 10-mile course for the Cleveland Pneumatic Aerial Trophy.

The first prize of $3,750 for the Trophy Race was won by Maude Irving Tait of Springfield, Massachusetts. She was a flier whose name and reputation stood high in aviation annals during the early 1930s. For a time she was New England governor of the Ninety Nines and with Amelia Earhart held a place on the contest committee of the National Aeronautical Association.

Neither she nor anyone else that year, however, equaled Mae Haizlip in high finance. At that one racing meet, including the second prize paid by the derby, she won $7,550. As if to warn her, however, not to play her luck too far, an accident befell the pilot the very next month. She had gone to a small racing meet at Clarksville, Tennessee, in October. As she took off for one of the events, her plane crashed. Escaping without vital injuries, she suffered a

Underwood & Underwood

BLANCHE WILCOX NOYES, one of the leading pilots in the Bureau of Air Commerce campaign for air-marking, and copilot and covictor with Louise Thaden in the famous 1936 Bendix Race.

MAE HAZLIP, record-winning pilot and first woman to be traffic manager for a major airline.

"Acme"

broken ankle and other hurts which kept her out of flying for some time. When the National Air Race period came around again, however, in September 1932, Mae was once more right up in the front ranks.

James Haizlip was there waiting for her. He had just made a transcontinental record, and the fast plane which he had flown was now turned over by him to his wife. It was a Wedell-Williams monoplane with a 540-horsepower supercharged Wasp engine made by Pratt and Whitney. To the National Air Races at Cleveland went the plane, the two Haizlip pilots, and eleven-year-old Hayes, who had a grandstand seat for the events.

"What I'd like to do," announced the ambitious Mae, "is to beat Ruth Nichols' speed record." In April 1931 Ruth had driven her plane at 210.704 miles an hour.

Climbing into the monoplane, Mrs. Haizlip zoomed over the Cleveland course at an average rate of 252.226 miles an hour. This was a new high record for women.

Amid the excitement she created, a friend rushed up to the little Haizlip boy in the grandstand. "That's your mother they're cheering for, Hayes!"

"Yeah, I know," he said calmly. "That speed's all right, but Dad can go faster."

Mae Haizlip, perched on the struts of a plane and chatting with reporters and fellow fliers, burst out laughing when she heard of her son's comment. "As a matter of fact," she remarked, "one doesn't get as much feeling of speed as one would suppose."

Then, turning to a newspaperman, she said seriously, "I think these trials are more than sporting events. We have a laboratory of speed here at the races which certainly stimulates the making of better planes and engines for transportation."

The next event of the meet which Mrs. Haizlip entered was the Aerol Trophy Race. Unfortunately the weather, beginning with murky fog, changed to a downpour during the fourth lap. The judges couldn't see. Neither could the fliers. Jokers declared that the winner could only be picked by flipping a coin. When final decision awarded Mae Haizlip second place, she declared she was cursed with a second-place jinx.

Next year, nevertheless, she was destined to break the jinx, for 1933 was a great year for aviation. Wiley Post made the first solo flight around the world in just less than eight days. Ruth Nichols was doing tricky flying abroad. Roscoe Turner won the Bendix Race across the American continent, and Amelia Earhart set a new transcontinental record for women of 17 hours, 7 minutes, and 30 seconds.

Two years before this Amelia had demonstrated a new craft on which engineers of several nations had been working for many years. It was the autogiro. This plane has rotary wings above fuselage and cockpit which offer such a powerful lift that the ship cannot only rise straight up from the ground without necessity of a long runway, but remain almost stationary in the air. The mechanism was not perfected until 1938. But in 1933 many improvements had been made, and Blanche Noyes was asked by the Standard Oil Company of Ohio to demonstrate the new Pitcairn model. Blanche had so mastered its mechanism that she equaled the far more experienced Miss Earhart's record of 150 hours' flying time.

In addition to varied activity in the flying world that year Mrs. Noyes became one of the pioneers in a new field. She was asked by Phoebe Omlie to join a group of women including Louise Thaden in a campaign to persuade the federal Government to take up the air-marking of towns.and cities. This undertaking, which was deeply interesting to Blanche, had a profound effect upon her future.

The last year of Mae Haizlip's notable public performance was 1933. In the Aerol Trophy Race at Los Angeles she won first place and a $1,500 prize. True, it was a slow race and her time was far below her own record; but she bettered her speed considerably in the National Air Races held in Chicago a few weeks later. There she won the Walter E. Olsen Trophy Women's Race with an average of 191.11 miles. Upon these successes, which banished the threat of forever holding second place, the energetic flier was content to rest.

Yet the year wound up with a signal honor for her. A great aviation ball was held in Washington by the Aero Club. Climax of the glittering occasion was the presentation of the certificates of award to outstanding American fliers of the previous year, 1932. These were offered by the National Aeronautical Association and its president, Senator Hiram Bingham, delivered the awards. Two men and four women had been selected. Major James H. Doolittle, James G. Haizlip, Amelia Earhart, Frances Marsalis, Louise Thaden, and Mae Haizlip—such was the galaxy of talent.

Often these four young women had been photographed just before or after a flight. Now, instead of boyish flying suits and goggles, lovely evening frocks made them so glamorously feminine that their association with dangerous adventure seemed preposterous. As the pilots stepped forward and bowed to the applauding audience one after the other, the other three husbands doubtless envied James Haizlip. There he was right up on the platform actually sharing equal honors with his wife! It was a scene to gratify one of the old campaigners of woman-suffrage days.

If Mrs. Haizlip no longer appeared in contests after that, it was not because she had retired from action. Far from it. Her reputation as a flier made her eligible for an important commercial position. By the St. Louis, Louisville, Evansville, and Cincinnati Division of Columbia Airlines she was engaged as traffic manager. Never before on a major air line had a woman been entrusted with such responsibility. Her knowledge of planes and routes and her familiarity with flying problems were the special equipment needed for this work. Morever, she had what was perhaps a still more important asset. As one of the men associated with her put it, "Mae has personality plus!" Some three years later Mrs.

Haizlip went to Europe for a long period of residence and on her return settled in Tulsa, Oklahoma.

Some time before Mrs. Haizlip became traffic manager, Blanche Noyes was also entering a new phase of aviation. Her husband was teaching her instrument flying. Very few women had this instruction. Indeed, it was not until 1939 that a woman flier ever made a blind landing. That was Jacqueline Cochran's feat. Blanche was an apt pupil and was also deeply interested in navigation. She and Dewey told their friends they intended to take a flight around the world.

But it was not to be. Shortly after Blanche won the Leeds Trophy in 1934 the fatality ever poised above a flier's head struck down her husband. For some time he had been air pilot on the Fort-Worth-to-Cleveland run. One night his plane crashed, and America lost one of its gifted aviators. To Blanche the tragedy meant that the radiant center of existence was gone.

It was only the self-control acquired by every good flier that carried her through the profoundly moving drama of her husband's funeral. The ceremony was held at the hangar of the Ninety Nines on Long Island. As the ambulance plane from the West, bearing the body of the dead flier, neared Harrisburg, it was met by three Army and three Navy planes. In exact formation the impressive cortege swept down upon the field.

There, supporting their comrade through the ordeal of the testimonial service, were Margaret Cooper and Amelia

Earhart. In the lonely months which followed Blanche was often asked to be Amelia's guest. The Ninety Nines put imaginative and constructive sympathy into an effort to lure Blanche back into action.

The organization asked her to visit the main cities of Ohio and Michigan to arouse interest in an all-women's cruise to the National Air Races of 1935. The contest board of the National Aeronautical Association had a committee on women's contests. The women's cruise was strongly advocated by the committee as a means of spreading greater interest in the meet. Mrs. Noyes bravely accepted the commission.

Perhaps it was partly because of the success with which this charming woman conducted the tour that she received a Government appointment. Eugene Vidal was now director of Air Commerce in the Bureau of the U.S. Department of Commerce. Vidal was a great figure in air circles. Starting as a West Point man, an all-American football player, and an Olympic champion, he had been an Army flier and engineer. In 1929 he had been a pioneer in transcontinental air transportation. Later he had organized an hourly service between New York and Washington, which had proved that a passenger service could be profitable. This official had taken up with enthusiasm the plan of air-marking cities in which Mrs. Omlie and Louise Thaden had been so deeply interested.

Mrs. Omlie had been in Washington for a year with the National Advisory Committee for Aeronautics. Louise

Thaden, Helen McCloskey, and Helen Richey were already veteran air-marking pilots. Therefore, it was with the anticipation of joining old friends that in August 1936 Blanche Noyes, the newly appointed member of the bureau staff, went to live in the capital. Hardly had she taken up her duties, however, when the shocking news of Captain Vernon Omlie's death brought back afresh her own irreparable grief.

Troubled about her friend's state of mind, Mrs. Thaden went with her on her first air-marking trip for the bureau. It took them to Texas. Out of the blue one day Louise received a thrilling message from an aircraft company which she had served for many years as demonstrator. She was offered a Beechcraft plane for entry in the Bendix Transcontinental Race. As she weighed the fascinating possibility, she decided that her companion might get a new lease on life from sharing the excitement of the contest.

Rushing into the hotel room occupied by Mrs. Noyes, she proposed the plan. "I want you for my copilot and navigator, Blanche. Say you'll go and I'll wire my acceptance to Beechcraft."

In the light of sudden interest the lovely face of the young flier lost some of its sadness. It took little persuasion to induce her to accept. Then followed days of swift and strenuous action. Wires were sent in all directions. Eugene Vidal gave the pilots leave from Government duty to enter the race and sent his blessing. The Beechcraft Company completed arrangements, and the two contestants raced up to New York. There for two mad days they went over the ship with the mechanics. At last all was ready, and just in time.

At four-thirty in the morning of September 4 the young women set off on the long trek. The weather was not good and it grew steadily worse. Then the radio failed. But, aided by Blanche's knowledge of blind flying, they held to the course and reached Los Angeles first of all the contestants. They themselves could hardly believe the amazing news shouted at them by the crowd at Mines Field.

On a great wave of national glory, the two air-marking pilots sailed back to Washington. Louise Thaden, however, did not pause there long. An offer from the Beechcraft Company swept her back into the field of commercial piloting and demonstrating. As Blanche took up her new duties, she found herself, as co-victor of the Bendix Race, welcomed more enthusiastically than ever at the towns she visited on air-marking pilgrimages.

Much of this work had been accomplished through funds delegated to cities and towns by the Works Progress Administration. When, therefore, the appropriations for such expenditures were cut, it was difficult for the bureau to proceed. Some 9,582 towns remained unmarked. Mrs. Noyes was assigned the task of persuading the Chambers of Commerce and other local associations to co-operate by investing their own money in painting town names for aerial guidance. It meant almost constant travel for her and the exercise of all the arts of diplomacy. Only an experienced flier with established prestige could have fulfilled these obligations.

Gradually all the other air-marking pilots left the Bureau of Air Commerce. But Mrs. Noyes remained. One by one the unmarked towns flaunted their names on roof tops. When that was accomplished, the pilot had to make the rounds again to see that whenever the gigantic lettering grew dim it was repainted.

In wartime, when planes move constantly across the country, such guides are of supreme importance. Many a newly made Army pilot, trying out his wings for the first time, must have blessed that giant lettering. A single downward glance told him just where he was and where he ought to go next. How many of these pilots would ever guess that a delightful woman is responsible for the efficiency of a system that guides them safely on their way?

Amelia Earhart

SHE DRAMATIZED FLYING

☆

ON JUNE 5, 1928, a shining Fokker plane arrived in a tiny fishing village in Newfoundland. On June 16 that plane was still there. Meanwhile, the inhabitants of Trepassey, as the village is called, were kept in a constant state of excitement. Time and again the rumor, "She's going to set off today," had brought dozens of people to the harbor. Each time, however, they had gone home with dejected faces.

But there was no one in Trepassey under greater tension than a certain young woman who had appeared simultaneously with the Fokker. Her tall figure, slim hipped as that of a boy, was seldom at rest. Even when she sat, the beautiful long, tapering fingers strummed impatiently on the table or desk. Always the wide-set gray eyes under that mop of blonde hair were puzzled and sometimes they were gloomy. Would the wind never be right to take off? Would there be any more trouble with the ship? Was she going to lose her big chance to be the first woman ever to cross the Atlantic in a plane? So wondered Amelia Earhart for almost two weeks.

Not the least of her anxieties concerned the pilot who was to fly her. He was a genius at flying, this Wilmer Stultz, but he often chose the avenue of escape so popular with genius. When things got on his nerves, he drank to excess. Certainly he showed the effects of many bumpers on the morning of June 17. And as Amelia took in the mottled cheeks and heavy eyes her heart sank. Cross the Atlantic with Stultz in this condition? Did she dare?

Yet this morning at last the weather was good. Feeling that it was a case of now or never, she took the desperate chance. She almost dragged Stultz to the plane. Shortly afterward there was a flash of silver wings against the Newfoundland sky. They took upward and onward, an orange-red body. Like some gorgeous tropical insect, the Fokker set off for England.

Those precarious hours when fog and rain blotted out the sea below, when the boyish young woman crouched in the cabin and wondered whether she should throw overboard the bottle which her pilot had smuggled into the ship; the victory of the pilot's flying instinct over alcohol; his magnificent feat of flying blind for 2,246 miles at an average speed of 113 miles an hour; the landing at Wales; the first glimpses of Great Britain; that evening at the Embassy Club when the current glamour boy of the world, then Prince of Wales, danced with her until the musicians wondered, "Isn't he ever going to signal to us to stop?";

the triumphant return to America—all those moments which lifted an unknown social worker in Boston to a peak of fame are familiar to most of the reading public. Not only were they described by Miss Earhart herself in her book, *The Fun of It,* but they are sketched by her husband, George Palmer Putnam, in his excellent biography, *Soaring Wings.*

The truth is that of all women aviators who ever lived Amelia Earhart has been the most publicized. Newspapers and magazines gave extravagant space to everything she did. Her lecture tours brought her before millions. Dubbed "Lady Lindy" and "First Lady of the Air," she was mobbed by crowds of curious folk. Even her last fatal journey is recorded in her husband's book, *Last Flight,* from the notes which she left behind her. Today, therefore, it is very hard to decide whether Amelia Earhart was the greatest woman flier of her day or merely the greatest personality of her sex who ever flew.

The strange part about the publicity which beat upon her ever after her flight to England in 1928 is that she herself had little to do with it. That she was essentially a modest person, always anxious to give the other fellow his due, is witnessed by both her words and deeds. For example, when President Coolidge wired her his congratulations after that first crossing of the Atlantic, her answer was that all the credit belonged to the pilot. Indeed, she is quoted as saying to a friend about that precarious journey, "Why bring that up? I was only a cuckoo in the nest."

About a career which has already been so completely revealed there is little left to say. For this reason we shall content ourselves here with high lighting a few spots in her meteoric life. Those who are curious to know more about Amelia Earhart must turn to both her own books and those of her husband, George Palmer Putnam.

She was born in Atchison, Kansas, in 1898, of sturdy American stock. Her father and paternal grandfather were both Lutheran ministers, but her maternal grandfather was more opulent. He was Judge Otis, distinguished member of Atchison's "leading family." After a happy childhood, in which she showed always that zest for living which was to distinguish her later years, she proceeded from various high schools to that citadel of Victorian young ladyhood, the finishing school at Ogontz, Pennsylvania. After graduating here, she listened to the call of World War I and went to Canada to become a nurse's aid in a military hospital.

At the end of the war we find her enrolled at Columbia University as a premedical student. However, a year changed her plans and she went to California to join her father, who had recently moved to that state. Here were laid the foundations for her future. She began lessons in aviation.

Only a little more than twenty years ago! Yet today that air field in southern California where the tow-headed Amelia first began to play about seems as remote as the days when knighthood was in flower. A desolate patch of ground where rested the "crates" in various stages of decrepitude.

AMELIA EARHART PUTNAM, widely known and loved as "A.E." and "First Lady of the Air," in a characteristic pose shortly before she took off on her ill-fated last flight.

A trick of putting together your own little jalopy and hoping that it would hold together. A daredevil race with the air and with your machine which trusted, not to safety devices, but to Lady Luck.

Typical of those flying days was Amelia's first instructor. It was a woman. Her name was Neta Snooks and she was the first of her sex ever to graduate from the Curtiss School of Aviation. She had cropped red hair, numerous freckles smeared over with grease and dust, and she capped these male assets with an old pair of dirty overalls. It was no wonder that some of her pupils said meekly, "Yes sir," when "Snookie" told them anything. But she knew her job, and when she turned Amelia over to an ex-Army flier for further instruction the girl was well prepared.

In May 1923 she was granted her pilot's license by the N.A.A. and during that same year we learn from her book, *20 Hrs., 40 Min.,* that she made two altitude attempts. One carried her to about 14,000 feet and the second to 12,000 feet. It was with this excellent record that she went to Boston, Massachusetts. Here her duties as a settlement worker at Dennison House left her some free time for her favorite sport. She became vice-president of the Boston chapter of the N.A.A. and also one of the five incorporators of the Dennison Aircraft Corporation. Thus we find her at the time she was picked by a group of interested persons to become the first woman passenger in a plane trip across the Atlantic.

This record denies emphatically one persistent legend.

The girl who was ferried from Newfoundland to Great Britain was no untrained convert to aviation. She was already an excellent flier. However, it was only when she returned to America that she began intensive work. In April 1929 she obtained the coveted transport license then held by only three other women fliers—Ruth Nichols, Phoebe Omlie, and Lady Mary Heath. Just two months later we hear of her being involved in one of the most argumentative little affairs in the history of aviation. This was the famous Powder-puff Derby from Santa Monica to Cleveland upon which we have already touched.

It had started seriously enough—the proposal for that race between women pilots. But before the plan got well under way it began to look like the libretto of a musical comedy. The lady pilots argued with each other and the men pilots began to jibe, "Why doesn't each of you take a man with you?" At this point *Time* records that "Amelia Earhart barked long and furiously."

The year 1931 was destined to become a banner one for Amelia Earhart. She was engaged by *Cosmopolitan* magazine as aviation editor. Transcontinental Air Transport appointed her assistant to the general traffic manager. The lectures which always supplied her with a certain revenue were now well under way. And in February she married the man who had been instrumental in choosing her for the flight from Newfoundland to England. He was George Palmer Putnam, publisher and writer. Known as "Gip" by his friends, he was regarded by many as the real impresario of this prima donna of the air. Certainly he did nothing to dim the limelight which always played upon "A. E.," as she was now almost universally called.

The year 1931 was marked by another distinguished experiment. She had always been interested in the autogiro, that plane with the revolving blades on its top. This was the type of flying machine which had haunted Leonardo da Vinci in his Renaissance workshops. But only about this time was it coming into its own. Because it can rise and descend vertically; because it offers a solution of the dream of landing on a roof top; because it can remain stationary above a given location—that was why this variation of the ordinary machine always fascinated her. She became the first woman ever to make an autogiro ascent, and on her second attempt at an altitude record she bested anything yet done by man himself. She rose to 19,000 feet. Later on, too, she made a transcontinental flight in one of the Windmills.

But her greatest "first" was to come the following year. While acknowledging generously the flying genius of Stultz on her first trip across the Atlantic she had perhaps been stung by certain unfriendly comments. "Backseat driver" and "excess baggage"—these were not descriptions which she intended to endure for the rest of her life. The plan for piloting herself to England had long been close to her heart. On May 20, 1932, five years after Lindbergh's epic flight, she set out alone from New York in a

Vega monoplane with a 500-horsepower Pratt and Whitney engine.

It was a journey even more perilous than her first. On that dark night, when the moon hid behind the clouds, lightning stabbed at the little red-and-gold plane and a wind shook it as a terrier does a rat. In the midst of the storm the lonely girl at the controls noticed with horror that something had happened to her altimeter. She could not tell how far she was above the mountainous whitecaps. Instinctively she began to climb. If she could only get above the clouds!

Half an hour of that climb and—yes, her wings were growing heavy. There was ice upon them. The plane went into a spin, and not until she was uncomfortably close to the hungry breakers below could she right it. An interlude of peace and then, just as she glimpsed a single star between the clouds, she saw that something was the matter with a manifold ring of her engine. It was trailing flames. A weld had broken.

How long would it take to burn through? Whenever that moment came, she knew full well it would be the end. Yet what could she do? Impossible to turn back. She must go on, go on with that single gnawing thought, "How long can it last?" When the dawn came at last she could hardly believe in the miracle. She was still alive.

But even those first wan streaks in the sky could not bring assurance. Still the flames lashed from the broken weld and now the weakening metal rattled ominously.

Could she possibly reach land before——? Wondering this, she turned on the reserve tanks and saw that she had a leaky gauge. Her heart stood still at this, the third threat to that lonely night. Land, land. It had to be. No time to waste. Dropping down along the coast of Ireland, she came to rest in an emerald pasture where cows chewing their cuds and fixing upon her a solemn curiosity seemed to say, "Why should you ever want to go up there when you can stay here on nice, safe, comfortable earth?"

It had happened. A woman had flown the Atlantic. And she was an American woman. One wonders if the ghost of another American woman was there to welcome her to British soil. Did beautiful Harriet Quimby in her mauve-colored satin suit whisper, "Well done, sister"? and then add wistfully, "Ah, if only I could have flown your plane!" She had a right to be present. For just twenty years before Harriet Quimby had made history by crossing the Channel —the first woman in the world to do so.

Not only had A. E. won her title. By her flying time of 13 hours and 30 minutes she had beaten the transatlantic record. Moreover, the 2,026.5 miles she had covered represented the longest nonstop flight ever taken by a woman. Small wonder that during the weeks she remained in Europe all the laurel wreaths were dusted and placed on that mop of blonde hair.

In England the late King George and Queen Mary sent their congratulations through Great Britain's Minister for Air, and Lady Astor wired, "Come to us and I will lend you a nightgown." In Paris she was the first woman ever to be received officially by the French Senate and there, too, she was awarded the Cross of the Knight of the Legion of Honor. In Brussels she was entertained *chez famille* by the late King Leopold and his consort, and when she left she was in possession of the Cross of the Chevalier of the Order of Leopold. Even Italy almost forgave her for the fact that she had strayed from that polite concentration camp, the home, where Latins have always liked to keep their women. Later she was to receive a decoration from the Italian government.

It was all enough to turn the head of the average man or woman. That A. E.'s topknot remained in the conventional place through all this adulation is indicated by the way in which she summed up her hazardous feat. "Literally hundreds of persons have crossed the Atlantic Ocean," she wrote, "including airplane and dirigible flights over the North and South Pacific, so that my flight can be of little aid to aviation as a science—except as it demonstrates the reliability of modern aircraft." This is thoroughly in line with the modesty which this publicized woman always displayed.

We may pass over the next few years with that many words. However, mention must be made of the feminine record, Class C, air-line distance, when she flew from Los Angeles to Newark, a distance of 2,447,728 miles, in 19 hours and 5 minutes. Using her Lockheed Vega monoplane with a Pratt and Whitney Wasp engine of 450 horsepower,

she thus surpassed by about 500 miles the record set by
Ruth Nichols. Subsequently she was to better even this
record by covering the same ground in 17 hours, 7 minutes,
and 30 seconds.

It is the year 1935. This was a year so brimming with
activity that the reporters' typewriters had to work over-
time. It began with her appointment to the Bureau of Air
Commerce, an honor which paid her one dollar a year.
Her special job was to demonstrate the new directional
radio apparatus and she did it well. Yet this post did not
interfere with more spectacular achievements. On March
23, 1935, she made the historic flight from Pearl Harbor to
Oakland, California, which added still another laurel leaf
to her crown of victory. We shall let *Time* magazine tell the
story.

A week-end recess in the Hauptmann trial cleared the front
pages of the papers for a good spot news story, and her flight
filled it. A. E. likes to say that she flies for "the fun of it." Last
week her fun consisted of flying blind through fog while she
listened to musical broadcasts and exchanged witticisms with
her husband by radio. Some eighteen hours after the start of
the 2,400-mile flight she landed safely at Oakland, California.
Back in Honolulu, Husband Putnam made good copy by saying,
"Myself, I'd rather have a baby."

Let us add a few details to this terse and salty record.
In that Lockheed Vega equipped with a supercharged Pratt
and Whitney S1D-1 Wasp and a variable pitch Hamilton

Standard Steel propeller, she made the trip in 18 hours, 16 minutes. About 525 gallons of gas and 35 gallons of oil gave the ship a cruising radius of more than 3,000 miles. Throughout the trip she flew at an average height of 8,000 feet and, though she encountered many rain squalls, cloud banks, and fogs, there were no severe storms. In spite of head winds, she averaged more than 140 miles an hour. Remember, too, that no radio beams spared her from constant calculation. She navigated by dead reckoning supplemented by position fixes from ships and shore radio stations.

Behind such details lies one stark fact. It was the first westward crossing ever made and A. E. had done it. In recognition of the achievement she ranked as outstanding woman flier of the United States. Also, she became co-holder of the title, world's leading woman pilot. Jean Batten, the New Zealand girl who flew the South Atlantic the same year, halved the honor with her.

After this stunning achievement most people would have said, "Well, I've done my bit for 1935." But A. E. was never one to sit on her laurels. To her those laurels were air cushions always in need of filling. So a trifle more than three months after her Hawaiian flight she rode the skies between Burbank and Mexico City.

In order to finance this flight the Mexican government, which had proposed it as an expression of good will, issued nearly a thousand stamps printed with the words, Amelia Earhart, Flight of Good Will, Mexico, 1935. These stamps,

AMELIA EARHART PUTNAM arriving at Oakland after her solo flight from Honolulu.

The Lockheed monoplane in which she made her famous solo flight across the Atlantic in 1932.

inscribed, of course, in Spanish, are today a treasure in any collector's album. At the time they accomplished their purpose. The cost, always a major problem with the aviator, was settled. But there were other hurdles. How about the terrible hazards? "Don't do it, A. E.," warned the late Wiley Post, always a great friend of hers. "It's too dangerous." He was only one of those whose advice she scorned.

She left Burbank at about one o'clock in the morning of April 20, 1935, sailing over a Mt. Wilson silvered by the moonlight. Mountains—deserts—water. They were hers again by moon and dawn and day's full light. It was only in the afternoon that jubilance turned to dismay. Over the Mexican state of Hidalgo both radio and compass went sour. What was more, a bug flew into her eye to prevent her from reading the map. About forty miles from Mexico City she was obliged to come down. Again it was in a pasture—but this time not the lush green of her Irish sanctuary. This was a lake basin covered with cactus and prickly pear.

"What is the affinity between me and pastures?" she asked laughingly afterward. To this query somebody replied, "You're just making good on the old line, 'Fresh woods and pastures new.'" Still another wag spoke of her as the pasteurized flier.

Her brief delay did little to spoil her record for those 1,700 miles. She had covered them in 13 hours, 32 minutes. Mexico City, which almost a quarter of a century before had welcomed Mathilde Moisant, first woman to fly over that mountain-rimmed capital, greeted her with wild Latin enthusiasm. President Cardenas made her a flowery speech, and his wife handed her a medal from 200,000 Latin-American women. And when she came away it was with the sombrero-topped costume of a Mexican cowboy. It had been presented to her at a huge concert given in her honor.

By her nonstop flight from Mexico City to Newark she achieved another record for both men and women. She made the 2,100 miles in 14 hours, 19 seconds. Back in New York she was greeted with the usual fanfare of newsmen. A few days later the air-minded Mayor LaGuardia presented her with the city's medal for "distinguished and exceptional public service." Later on in the year Carrie Chapman Catt, pioneer suffrage leader, named her as one of America's ten outstanding women.

It was during the latter part of fruitful 1935 that A. E. found herself in a brand-new niche. She became a visiting faculty member of Purdue University. The "visit" represented about a month out of the college year, and during that time she was supposed to address the 800 coeds on vocational opportunities. More than that. This famous Hoosier college with its student body of 6,000 was at that time the only educational institution in the country which could boast its own airfield equipped for day and night flying. Her eyes glowed when she was told she was to be adviser in aeronautics.

They glowed even more brightly the next year when

Purdue announced that a fund of $50,000 had been sub-
scribed to be known as the Amelia Earhart Fund for Aeronautical Research. The fund would provide a Lockheed
Electra with every known appliance—deicing appliances,
robot pilot, two-way radio telephone, et cetera. It would be
the first "flying laboratory" the world had ever seen, and in
it A. E., working always in close touch with Purdue, would
test all the human reactions to flying. Diet, fatigue, altitude
—what did they and other elements do?

"It's simply elegant," she would crow in the July of 1936
after she had put the new "flying laboratory" through its
paces in the air. And the mechanics who always gathered
around on the California airfield to watch her land agreed
with her.

Indeed, Purdue University had proved a fairy godmother. How otherwise could she have possessed this $50,-
000 Lockheed Electra? A ten-passenger, two-pilot transport
plane of the type in use on various air lines, it had a 55-
foot wingspread. It was all of metal and its two 500-horse-
power Wasp engines gave it a normal cruising speed of
190 miles per hour. Did doom lurk in this magic gift?
How could A. E. have any prescience of the future as she
gloated over the beautiful shining ship? She knew only that
the Electra at last made possible an old cherished dream.
She was going to fly around the globe.

Not, however, as Wiley Post and others had done. They
had chosen the short northern route, a distance of something more than 15,000 miles. But A. E.'s course—that was

to be as near as possible to the equator. It was to cover 27,000 miles. From California to Honolulu and thence to Howland Island, a mere speck in the South Pacific. Afterward—New Guinea, Darwin, India, Dakar, Natal, Mexico, home. For the benefit of newshawks her long, tapering fingers traced on a miniature globe these stations of her prospective journey, a journey which she was to reverse. She did not know, of course, that she was pointing out many of the spots on the earth's surface which only a few years later would mean life or death for her country.

It was on June 1, 1937, that the red Electra sailed out into a Florida dawn. In spite of the early hour, a large crowd had assembled to bid godspeed to "the first lady of the air." It was the last American throng to see the slim figure, the wide-set gray eyes, and the gay smile of Amelia Earhart. For a month messages were received from the plane. But they ended abruptly July 2, 1937:

"Circling . . . cannot see island . . . gas running low."

With these words sent from the vicinity of Howland Island the woman who for nine years had made aviation history vanished from our world. For days afterward that world refused to give up hope. Surely she who had come through so many perilous adventures would emerge triumphantly from this. But at last we had to accept the truth. This time A. E. had found no sanctuary of earth. This time the only pasture was the vast expanse of the South Pacific. There under those same waters where her countrymen are now battling a ruthless foe she and her fellow pilot, Captain Fred Noonan, undoubtedly perished. The beautiful shining ship which she had greeted as her heart's desire had lured her to the last great adventure.

And now, five years after A. E. was drawn down to the ocean's bed, what shall we say of the woman flier who did so much to dramatize aviation? Let us meet the challenge by saying that she was much more than a flier. "This Amelia," once wrote the late Will Rogers, "she would be great in any business, or in no business at all." These homely words sum up her personality. She was a many-sided woman to whom aviation represented merely the dominant interest of a life brimming with interests. She herself once wrote that she had had twenty-eight jobs in her life and that she wished they were a hundred.

Run through just a few of her accomplishments. On the lecture platform she developed real excellence of phrase and delivery. At Purdue she proved a stimulating teacher. She wrote well—with humor and a nice feeling for words which she might easily have developed into a profession. She was a magnificent driver, albeit traffic cops found her somewhat of a problem child. Occasionally she turned gardener, and once she tried her hand at clothes designing. She could even sew, and now and then an intimate friend would drop in to find her making a dress. "The fun of it" —that was her own favorite phrase for the motive which made her poke into so many paths of life.

She was not a pretty woman in any conventional sense of the word. And the adjective "handsome," so often applied to her, hardly took in the gaiety, the humor, and the real kindliness which made the charm of her face. As to the alleged resemblance between Colonel Lindbergh and herself, it was based only upon height and slimness, the mop of blonde hair, and the wide-set eyes which both possessed. Nobody who talked with them together for half a minute could miss the difference in temperament which made them look so utterly unlike. Whereas Lindbergh always held himself aloof from people unless they were proven friends, A. E. liked folks of all kinds.

And who were "folks"? They were the late King Leopold of Belgium, meeting her in unpolished shoes and unpressed trousers and talking to her so simply and earnestly about flying problems. They were the boys in the firehouse in Coatesville, Pennsylvania, who offered to let her drive their new engine. They were Eleanor Roosevelt, and the mechanics on the airfield. They were anybody who happened to be vital, good, unaffected. With this outgoing nature, is it strange that her friends were drawn from every walk of life? Or that the newsmen, who had always been so put off by Lindbergh's tight-lipped manner, adored this heroine of the sky?

Many people have spoken of her as an ardent feminist. Certainly she loathed the attitude "this is man's work," and "this is woman's work." It was aptitude and not sex which she felt should determine any chosen vocation. She loathed even more the idea that marriage should absolve a woman from any activity outside the home. Once when a newsman told her that the only smart women were those who get a man to keep them she flashed back at him with one of her most brilliant phrases, "I should hate to think of marriage," she said, "as a sort of cyclone cellar." In spite of her liberal opinions, however, the impression she made upon most people was one of great femininity.

Perhaps no other woman flier except Anne Lindbergh ever had such a poet's sense of the beauty of flying as A. E. To be among the clouds; to catch through their rifts a single star or perhaps a slender, curving moon; to skim the snow-topped peaks and look down on saffron deserts unrolling like a parchment scroll—this was to her the elixir of life. And perhaps—here was one of the fascinating divisions in her character—this people-loving woman adored most those flights which she took in absolute solitude. For underneath all her friendliness there was something elusive, something uncapturable. "The girl in brown who walks alone"—this description of her given by one of the schools she attended touches off eloquently this other side of her character.

In his biography, *Soaring Wings,* her husband refers to the intense admiration she always felt for Anne Lindbergh. Perhaps there was something a bit wistful about this feeling. Perhaps she wished that she, too, could write so beautifully about the beauty she felt. For one must never forget that a love of poetry had possessed her heart when she was a child and never left it. Today some of her classmates in the premedical course at Columbia will tell you that they remember Amelia Earhart best as hunched over some volume of French verse.

She herself often scribbled bits of verse which she was entirely too shy to submit to any periodical. But one of her poems did find its way into print. It is called "Courage," and we give it here as a fitting epitaph for the high-hearted woman who dramatized so vividly that quality which every aviator must possess.

Courage is the price that life exacts for granting peace.
The soul that knows it not, knows no release
From little things;

Knows not the livid loneliness of fear
Nor mountain heights, where bitter joy can hear
The sound of wings.

How can life grant us boon of living, compensate
For dull gray ugliness and pregnant hate
Unless we dare

The soul's dominion? Each time we make a choice, we
*   pay*
With courage to behold resistless day
And count it fair.

Elinor Smith, Frances Harrell Marsalis, and Helen Richey

THREE COMETS OF THE TWENTIES

☆

TOWARD the end of the 1920s a new group of women fliers was making its appearance in American skies. Encouraged by the more experienced women aviators, profiting by improved planes and by general acceptance of women in aviation, they came swiftly into the foreground. Some of them carved out careers in which they continue to be active. But even those who played dramatic roles for only a brief time added their bit to aviation history. In establishing records, extending public interest in flying, testing planes, and teaching novices, they made an important contribution to aviation.

Youngest of the group and first to get her pilot's license was Elinor Smith, born and brought up in New York. In 1926, at the age of sixteen, this pretty blue-eyed child made her air debut. But the will to fly was aroused a long time before, during her first trip in a plane. Elinor's father, an actor, had his own little plane, and the moment he felt master of it, he invited his daughter for a ride.

It was a shining day with many clouds of brilliant white. Elinor felt as if she were rising to a city in the sky fashioned of wondrous citadels and towers. Curving around the floating structures, the little plane came face up to a tremendous white mountain of cloud. The girl gasped, "Oh! We're going to bump!" But they went right through. Mist whipped past the ship and gray fog curtains shut tight around it. Then all at once—out again on the other side of the mountain into dazzling sunlit space.

When they were on the landing field again, Elinor turned upon her father the look of one who has seen fairyland. "It's marvelous!" she murmured. "I'm going to learn to fly, too."

In 1928, aged eighteen, Elinor had her transport pilot's license. In September of that year she headed skyward from Curtiss Field, determined to establish an official altitude record for women. With a barograph properly sealed, she sped upward. Her altimeter broke. She grew cold and dizzy. Yet she kept on climbing until the gas gauge registered empty. Down she sailed then with two gallons of gas in the reserve tank and not a drop left by the time she taxied to a stop. The reading of the barograph was 11,663 feet. Others had flown higher, but without official registration of altitude. Therefore, this daring youngster had actually set a world record.

That she was hardly mature enough for the intoxication of success was shown by a thoughtless trick she played a month later. Forgetful of everything except the wish to

satisfy an expectant public, she set out one Sunday afternoon to the East River and skimmed under four of the
great bridges which span it. True, Elinor demonstrated
steady nerves and great precision of control. But such ig-
norance of law and disregard for the safety of people on
river craft displeased the judge before whom she was
brought.

Nevertheless, the mildness of the sentence he passed was
some comfort to all who held dear the reputation of women
fliers. "You are grounded for fifteen days," the judge said.

Goaded by the ambition to be taken seriously, Elinor
planned to beat the record for endurance flying. A new
one had been set on New Year's Day, 1929, out in California by a clever young aviatrix named Bobbie Trout.
She had remained in the air 12 hours, 1 minute. Elinor
meant to stay up longer. The flight, however, was delayed
by that bane of competitive aviation, bad weather. After
waiting days and days, the girl decided one evening that,
since reports gave no hope of clear skies next day, she
would go to a dance. Tumbling into bed that night at a
late hour, she was deep in dreams when the telephone
rang. The Mitchel Field official was saying, "Good weather
at last."

Then in the dawn's early light the butterfly changed into
an eaglet. Knowing that her father would be afraid to have
her fly after so little sleep, the girl slipped out of the house
in secret. By noon, with every detail checked, she was rising
into the air. Her plane was a Brunner-Winkle with a 90-

horsepower motor. Hour after hour it wove a tireless pattern high over the field. The deadly grind was relieved in the late afternoon by the arrival of her father's plane close beside her. On the fuselage Mr. Smith had chalked the information that by eleven o'clock there would be moonlight. Never was that soft illumination more welcome. For suddenly the lights on Elinor's instrument board went out. Clock and dials were blank.

Not far away the girl saw the broad beam of a searchlight. It came from the radio station WEAF at Bellmore, Long Island. Instantly she headed for the beam, banked into it, and let it flood over her board. With a thrill of joy she saw it was nearly one o'clock. The record was broken! Returning to Mitchel Field, she circled low and fired her little signal pistol. Presently came an answering blink of lights. This was official assurance that she had established a new record and that all was ready for her descent. Climbing out of her plane, she was greeted by the news that she had advanced Bobbie Trout's time by an hour and fifteen minutes. There was an instant of happy triumph. Then the endurance flier suddenly became a very tired girl, glad to be driven home by her parents and put to bed.

A little more than a year after this, on March 10, 1930, Elinor made a new altitude record of 27,419 feet. That superb success gave her immediate fame and prestige. She spoke on the radio, talked before various organizations, and demonstrated planes for the Curtiss Company. Meanwhile, she added to her reputation another endurance flight which pushed the record up to 26 hours, 21 minutes, and 32 seconds.

For a year Elinor Smith's altitude record stood unchallenged. Then Ruth Nichols smudged it out by going up more than 1,300 feet higher. Determined to snatch back the laurel crown, Elinor set out for another climb into the upper air. She reached 25,000 feet. Suddenly the motor stopped. Horror-stricken, the girl, with lips closed firmly around the oxygen tube, tried to discover the trouble. She twisted about, searching with frantic eyes and hands. Suddenly everything began to swim before her eyes. Her head fell back. She was in a dead faint. In her wild effort to get the motor started she had slipped a screw in the oxygen tank and cut off her supply.

When she opened her eyes again, they were fixed on the dial. What had happened? With a shock she saw the plane had dropped 20,000 feet. The motor was still dead. Landing on Mitchel Field was impossible. As the plane circled lower and lower, she saw a field beneath her and cut toward it over a stone wall. In front of her was a big tree and a deep ditch beside it. With lightning movements she saved herself from death. She cut the switch, snatched off her goggles, and clamped on the brakes in such a way as to turn the plane over before it struck the tree. Out of the wreckage she crawled unhurt.

Two weeks later, on April 10, 1931, the plucky girl made another attempt to better the altitude record of Miss Nichols. She used a Bellanca Skyrocket monoplane, and when she landed the altimeter showed a climb of 32,500 feet. Bellanca himself was on the field to congratulate her. Reporters sped to send the news by telephone and telegraph all over the country. With her unfailing generosity Ruth Nichols sent a warm wire of congratulations. Then all this glory blew away like smoke. The sealed barograph, calibrated at the Bureau of Standards in Washington, showed an altitude of less than 25,000 feet. The altimeter had been completely wrong.

For a short time after this crushing disappointment Elinor continued active in aviation. Then she married. The happy oblivion of domesticity engulfed her from then on, but her real achievements remain a valuable legacy to the women fliers she left in the field.

One of the observers of Elinor Smith's early flights was bright-haired, hazel-eyed Frances Harrell. She was then an enthusiastic novice just arrived from Texas. Left a small legacy, the girl had immediately given up her positon as credit manager of a store in Houston and had come North to enroll as a student at Roosevelt Field. From there she went to the Curtiss Flying School at Mineola.

On a certain afternoon the girl, clad in an old overall, was working with one of the mechanics. Suddenly she looked up at the sound of her name on the lips of her instructor. "Here is Frances Harrell, one of our most promising pupils. You might like to talk to her."

Unconcernedly Frances scrambled to her feet to be introduced to a magazine writer who was getting copy on the

FRANCES MARSALIS and HELEN RICHEY at the completion of their record-breaking refueling endurance flight in 1933.

"Acme"

ELINOR SMITH, who established an altitude flight while still in her teens.

flying school. Tossing back her curly hair from a face smudged with black, she said, laughing, "I can't shake hands with these greasy paws, but I'll sure be glad to tell you anything you want to know."

The soft Southern voice and the radiant look contrasted both with the costume and the competent air possessed by the aviation student. She explained that she had been helping to tear down a motor and put it together again. In reply to a question put to her by the interviewer, she said, "Indeed, I do know what I want to try for. Soon as I get my commercial license, I want a job selling airplanes."

Before Frances won the license, however, she had won the heart of her instructor, William Marsalis. With his encouragement and coaching, she reached her goal in two years and by 1930 was a stunt flier for the Curtiss Exhibition Company. She soon became an expert at barrel rolling and looping the loop. Often she gave exhibitions of her own, and won a national reputation for thrilling air feats. At the air meets this light-hearted girl was the center of attraction. Both an excellent pilot and a "regular fellow," she became a prime favorite among both men and women fliers.

Small wonder, therefore, that when Viola Gentry was promoting an endurance flight, she chose Frances to share honors with the more experienced Louise Thaden. This exploit of refueling in the air and piloting the plane for 8 days, 4 hours, and 5 minutes was the first single achievement of importance by Frances Marsalis. The flight was made in August 1932, and the ensuing sensation was tremendous.

Dividing fame with Mrs. Thaden, Frances was presented with her to President Hoover. For months she was in great demand as a speaker, and her flying exhibitions drew larger crowds than ever. That year the Ninety Nines, of which she was a charter member, elected her governor of the New York and New Jersey districts.

In late December 1933 Frances made a second refueling endurance flight. This time her partner was Helen Richey, a Pennsylvania girl who was just winning her first laurels as a flier. All during the holidays, when other young women were going to parties and wrapping up presents, this determined pair was circling ceaselessly over the airport of Miami.

Imagine spending Christmas Day in the clouds! At dawn they were shouting over the motor's roar, "Merry Christmas, Helen!" "Merry Christmas, Frances!" At noon they were eating a turkey dinner dropped through the air by the refueling plane. With the meal came a little imitation Christmas tree with a dab of tinsel on it. In the afternoon they received mail and greetings, and in turn they dropped notes of acknowledgment over the airport. Thus, even in exile, they shared the world's celebration.

Frances even had to spend her birthday in the air. Of course she was showered with greetings. But discomfort and fatigue had reached a pitch too great for gaiety. Frances dropped a gloomy note saying, "I'm growing a year older every day I'm up here." At last, after 9 days, 21 hours, and 42 minutes of residence in the plane, the weary fliers could descend with a new record of endurance for pilots and plane.

It was a promising opening for 1934. Frances Marsalis made a first in each of two minor air races and was then chosen as special demonstrator of the Waco plane. This was the kind of work she loved, and the months sped by until August. On the fourth of that month women fliers assembled from every corner of the land at Dayton, Ohio. Backed by the Ninety Nines, they had staged the first National Air Meet for Women. Of course Frances was one of the entrants.

Many who had gathered at Dayton remember how spirited and self-confident was the lovable Texan with the hazel eyes. It is a heart-catching memory. For Frances Marsalis never finished the 50-mile race. In making a turn around a pylon, her light biplane was caught in the slip stream of another plane as it cut in very close. Apparently the young pilot could not regain control of the ship. It roared into a nose dive, struck the ground full force, and turned over a dozen times.

Strangely enough it was her copilot in the endurance flight who won that day. Undoubtedly Frances would have finished first. When she crashed, Helen Richey moved up from second place and crossed the line as winner. Perhaps no good fortune was ever attended with more tragedy, and the thrill of the Dayton Races was quenched in mourning. Escorted by Navy and Army planes, the dead flier was brought by air ambulance to Roosevelt Field where William

Marsalis, Ruth Nichols, and many other fliers were waiting. The funeral was held in a hangar banked with flowers, and the blossom-laden coffin of Frances Marsalis rested on the front of a plane such as she had used in her last flight.

When the shock of horror and grief had grown less, Helen Richey began to realize that her victory in the Dayton Races was of great importance to her career. She had had only four years of flying and yet here she was in the front ranks of aviation. In addition to the $1,000 prize, she received the Fairchild Trophy and a Sperry artificial horizon indicator. Of all the congratulations offered her none were so welcome as those of the heroine of a solo flight across the Atlantic a year and a half before. Amelia Earhart had become one of Helen's warmest friends in the Ninety Nines.

The girl's first friend in aviation, however, was her father. Dr. Joseph B. Richey, superintendent of schools in McKeesport, Pennsylvania, was not an educator for nothing. He recognized talent when he saw it, and not only paid for his daughter's flying lessons, but as soon as she got her private license he bought her a plane. That his confidence was well placed was proved by the girl's steady advance. A writer in *Collier's* said of Helen Richey:

There are women pilots with more air hours to their credit, but none who has flown more smartly or with saner purpose. She has barnstormed in her flight toward her present renown only because there was nothing else open to a bright young girl pilot.

From 1930, when she took her first solo trip in the air, to the victory at the Dayton Races in 1934, Helen had seized every opportunity for varied experience. Aside from the refueling endurance flight with Frances Marsalis, she had flown in the Sportsmen Pilots' Cruise to Miami and had entered the Liberty Treasure Hunt from St. Louis to New York. Moreover, at Pittsburgh she had successfully passed the test for a transport pilot's license.

Such were the preliminaries for the great opportunity which lay before her on December 13, 1934. She was the only woman to enter a contest with eight men for a job as copilot on the Central Airlines operating ships from Washington to Detroit. This was the period of the Administration's effort to deal with the depression through the N.R.A. and travel from the middle West to the capital was exceptionally heavy. The post on the air line was much coveted. Consequently there was a tremendous to-do when Helen Richey won the contest and secured the position as copilot. Newspapers heralded the announcement as meaning new recognition of the value of women fliers. The Ninety Nines, especially Amelia Earhart, showered their fellow member with congratulations.

Of course Helen herself knew that her success was regarded with a jaundiced eye by most male pilots. The quiet competence of her work was supported by a casual manner and an endearing grin. She asked no favors and treated herself neither to airs nor to outbursts of temperament. Three round trips a week was her schedule. Over the Allegheny Mountains, swept by tricky winds, and across the valleys of the middle East to Detroit she carried mail and passengers. Then back she flew to Washington, where she usually landed about three in the morning. Turning the ship over to waiting mechanics, she would slip into the lunchroom at the airport, sip a cup of coffee, and then go home to bed.

There were in Washington at that time five women fliers working for the Government. But Helen Richey, commercial air pilot, was news. On her off days she was often asked for interviews and, although she said as little as possible, many stories about her found their way into the journals.

"I'm keen about my work," she would say in a matter-of-fact voice, "and I think regular piloting is worth a lot more than record flights."

All her discretion, her skillful work, and the regularity of her schedule, however, could not offset her one great handicap. She was a woman in a man's world, and antagonism closed about her. It was announced that the pilots' union would not admit Miss Richey as a member. No woman could join. The voice of the union is very powerful. Its first effect was to draw from the U.S. Department of Commerce, which was actively promoting safety in air travel, the statement that it would not countenance Miss Richey's being sent out as copilot in bad weather.

Helen found her employers unwilling to act in conflict with these pronouncements. In vain she pointed to her record, asked if she had ever proved to be a sissy who demanded specially favored conditions. There was no way to liberalize the attitude of other pilots, and her only way out of the situation was to resign. Headlined in the journals, the story of Helen's resignation was reported far and wide. Most of the papers took the stand of one sympathetic commentator who said, "As every airwoman from Ruth Law to the youngest Ninety Niner has sadly learned, admitted skill is one thing, equality in aeronautical employment, quite another."

Women fliers publicly lamented the loss of Helen Richey to commercial aviation. Amelia Earhart issued a sharp protest against this flagrant sex discrimination. In answer to her attack on the Department of Commerce, Phoebe Omlie, technical adviser for Air Intelligence, came forward with a general defense of the department's work for aviation. Meanwhile, the storm center herself went sedately back to her family in McKeesport. She refused to be drawn into argument. Doubtless she was touched when almost immediately Clarence Chamberlin, defender and guide of women fliers, offered her a job on his air lines. Her refusal of it was due to the unfolding of another kind of opportunity.

Even before the end of 1935 the Bureau of Air Commerce in Washington appointed Helen Richey air-marking pilot. She was associated in this work with comrades of racing meets and of the Ninety Nines—Louise Thaden and Helen McCloskey. The next year Mrs. Blanche Noyes joined the bureau. Slowly and systematically the pilots were covering cities and towns all over the country and were getting co-operation in painting on roof tops and oil tanks the name of the locality in letters legible at 4,000 feet in the air. It was interesting and important work for aviation.

Hardly had Helen started on her new duties, however, when she was given a few weeks off to co-operate with the National Aeronautical Association. That organization was disturbed, because, as compared with other nations, the United States held third place in aviation records. A drive was on to stimulate American pilots to make forty new records at the earliest possible moment.

A prize of $100 was offered for the first record made after February 1, 1936. Helen Richey decided to try for a speed record in the light-plane class. Securing a leave of absence from the bureau, she borrowed a plane and flew to Langley Field at Hampton, Virginia. There she carefully tested out field and plane the day before the official flight. Here was certainly a case of the early bird getting the prize. The clock had just passed midnight and February 1 was but three minutes old when Helen took off. Buffeted by bitter winds, she drove over the 100-kilometer closed circuit at a speed of 77.224 miles per hour. This was more than top speed for Class C light airplanes, and when she landed at 12:55 A.M. the $100 prize was hers.

In May of that same year Helen was given leave to make another record. This time it was an altitude record for planes weighing less than 440 pounds. Then in September routine was once more interrupted. Amelia Earhart had in-

Frances Harrell Marsalis, nationally known stunt flier and joint holder of two endurance flight records.

Helen Richey, first woman co-pilot for a commercial airline.

vited the younger girl to serve as copilot in the Bendix Air Race from New York to Mines Field, Los Angeles. Joyously she accepted.

That was the race of bad weather, high winds, and many disasters. For hours the two young women had to struggle with a cabin door which had got loose and banged in the wind. They arrived safely, but had been so delayed that they came in fifth. Louise Thaden and Blanche Noyes won, and Laura Ingalls came in second. It was a great experience for Helen and certainly a great day for women fliers. The three pilots of the Bureau of Air Commerce returned to their tasks covered with glory.

In a few short years of strenuous effort and widening experience Helen Richey established herself professionally as a dependable flier. After a long period of service as airmarking pilot she applied for a license as flight instructor. The test was given by the Civil Aeronautics Authority at Roosevelt Field, Long Island. Helen was the first woman to pass the test and receive her license. The next year she added to her equipment as instructor a third-grade radio license. As the United States approached the war crisis, therefore, this young pilot was prepared for any service which might be required of her. She is an outstanding representative of the intelligent, sturdy, and well-trained women fliers of the nation.

Laura Ingalls

CLIPPED WINGS

☆

W<small>HILE</small> our American women were doing their best for the Stars and Stripes, women under Great Britain's Cross of St. George were not by any means loafing. Among those who attained great prominence on the other side of the water was a handsome Irishwoman, Lady Heath. This famous flier has written a book and in it she tries to convince a skeptical world that there's nothing easier than looping the loop. "Heaven bless you, why, it doesn't require any more spirit than to say, 'Please pass the butter.' "

Well, Lady Heath hasn't corrected the bad case of goose-flesh which most women get at the very thought of looping the loop. And it is noteworthy that she doesn't say a word about the difficulty of doing the outside loop. Reason enough for her silence, for this is a maneuver which makes even your veteran of the air shiver a bit. Why? Because, unlike the inside loop which pulls you in with your plane, the outside one pulls you out into space. That is why the performance of Dorothy Hesters, a flier from Portland, Oregon, dazzled the aviation world. For she smashed all world records for the outside loop.

The teacher of Dorothy Hesters was Tex Rankin, supreme in the field of instruction in stunt flying. Rankin also had under him at one time one of America's most celebrated women pilots. It was Laura Ingalls. But long before she sat at the feet of Rankin, Laura had been up to tricks. Indeed, from the moment she began to soar she began to loop. In May 1930, just after her graduation from an aviation school, she climbed up into the air and made 980 consecutive circles. The following August she broke both the men's and the women's record for the barrel roll. Her 714 rolls bettered by more than 297 the score of Dale Jackson, the male champion, and by 644 that of Betty Lund, the previous feminine champion.

However, Laura Ingalls is more than a human pinwheel, more than a mere stunt flier. She is one of the most distinguished pilots whom the United States has produced. Aside from Louise Thaden and from that three-time winner, Jacqueline Cochran, she is the only American ever chosen by the International League of Aviators as the world's outstanding woman flier. For it must be remembered that in 1935 Amelia Earhart was obliged to share the title with Jean Batten, the New Zealand expert. In that same year the Harmon Trophy, awarded by the league to the season's best aviator of either sex, went to a man. However, this prize, the most coveted in the whole field of sports aviation, did go successively to Louise Thaden, Laura Ingalls, and Jacqueline Cochran.

It is because she occupies such a prominent place in our history of aviation that the sentence passed upon her in February 1942 has such tragic meaning. Convicted of being in the pay of German agents without registration of her status, she is now paying the penalty for her offense. To plummet from the skies where she once ruled as a queen to a sordid and fixed plot of earth—here is a destiny more shattering than any other in the pages of flying.

However, since this book is intended to record the achievements of American women in the air, we must dismiss the personal fate of Laura Ingalls and deal only with her career as an aviator. To begin with, she was no Cinderella of the air. To her the plane did not mean any transformed pumpkin coach bearing her to wealth. She was born in New York City of socialite parents who sent her to private schools both here and abroad. When she came back to America after study in both Paris and Vienna she was an accomplished linguist. Also, she had devoted much time to the study of music.

It seemed natural enough that with such a background one of the arts should claim her. She decided to take up dancing and for a short time was engaged as a specialty dancer. However, the life of the theater did not satisfy her. She longed for some career which would not tie her down to one fixed routine. Flying—that seemed the solution, and in 1928 she enrolled at Roosevelt Field, Long Island. Subsequently she went to the Universal Aviation School at St. Louis and completed the course there in April 1930. This fact made her the first woman student in the United States to graduate from a Government-approved school. Also she became the first St. Louis graduate of her sex to apply for a transport license.

It was almost immediately after she got her license that she started those amazing loops and barrel rolls of which we have spoken. "She's not a flier—she's a wholly roller"— perhaps the words of a certain wag may have reached her ears. Perhaps she just got tired of doing fancy stitches in the air and had a yen for plain sewing. At all events the papers of October 9, 1930, brimmed with news about Laura. Flying a DH Moth open biplane with a four-cylinder Wright Gypsy motor, she had shattered the East-to-West record for women. She had winged from Roosevelt Field to the Grand Central Airport in an elapsed time of 30 hours, 27 minutes.

However, her laurels soon withered. No sooner had she completed a hazardous flight back to New York than she heard that her precious East-to-West record had been broken. Mrs. Keith-Miller, the English aviatrix, had clipped off almost five hours' time from those 30 hours, 27 minutes.

"A plane, a plane, my kingdom for a plane!" In 1931 she kept muttering this wish with a fair degree of consistency. For, although the depression which marked President Hoover's administration was now sapping the spirit of the average American citizen, the aviator mind kept ticking away, "What record shall I hit now?" The women's camp this year was certainly buzzing. Ruth Nichols, Edith E. McColl of Canada, and Elinor Smith were all talking about a transatlantic hop. It was natural that Laura Ingalls should wish to enter that contest. Unlike Elinor Smith and Edith McColl, she did get a ship. It was a beautiful Lockheed Express monoplane. Nevertheless, it never did take her across the Atlantic. For some reason she gave up the trip after the disastrous accident which befell Ruth Nichols. And not until the following year, when Amelia Earhart made her famous hop, did the dream which Katherine Stinson had cherished two decades before actually come true.

In 1932 the broad-shouldered, handsome man who had first dawned on the political horizon when Katherine Stinson and Ruth Law were the queens of the sky was elected President. Perhaps Laura Ingalls listened to Franklin Delano Roosevelt's inaugural speech—that stirring one in which he promised to "drive the money-changers from the temples." But certainly she could not have divined that some day her opposition to the foreign policies of this same President would seal her fate.

During the first year of Roosevelt's administration Laura Ingalls performed no achievement worthy of note. The second was equally blank. But in 1934 she succeeded in rubbing elbows with such news items as the latest doings of the N.R.A., the new unrest in Cuba, the Rasputin drama in the London courts. She became the first woman to make an aerial circuit of South and Central America. That tremendous flight of 17,000 miles took her over the regal, snowcapped Andes, and it was so perfectly executed that she met with not a single mishap. "I didn't even lose a handkerchief," she laughed afterward.

Here is an exact record of that momentous trip:

March 14, arrived at Talera, Peru, after a flight from Colón, Panama.

March 15, Las Palmas Airport, Lima, Peru.

March 29, arrived Santos, Brazil, after a 500-mile flight from Portalegre.

April 8, arrived Maceió, Alagoas, after seven-hour flight from Caravelas.

April 22, landed at Rancho Boyero Airport, Havana.

April 26, completed 17,000-mile solo flight through Central and South America, landing at Floyd Bennett Field eight weeks after take-off.

With this record it is not strange that she won the crowning honor of any aviator. It was then—in 1934—that she was chosen as the world's outstanding flier and that she received the Harmon Trophy.

Laura Ingalls always had the same effect on rumors that a piece of lighted punk has on a package of firecrackers. She made them go off and shoot in all directions. After she came back from South America the explosions were particularly violent. "She was going to enter the great London-to-Melbourne race scheduled in October?" "No, not at all.

She was going to fly the Pacific solo." "Now whatever made you think of that?" "She's going to fly the Atlantic solo and beat Earhart's time." It was thus the guesses raged through the world of aviation.

Much of the speculation arose from her new plane. It was being built for her behind closed doors out in Burbank, California—that Lockheed Orion monoplane. Report said that it was the fastest of all planes. And when it finally did emerge from the plant, people agreed that it justified all the advance publicity. It had a supercharged 550-horsepower Wasp engine. It could carry 640 gallons of gas. As for the equipment, this was princely. Retractable landing gear, controllable pitch propeller, flaps, Sperry gyropilot, Westport radio compass, a radio receiving set, and the other standard instruments—all were here. No doubt about it. This plane meant business.

But what business? Not until 1935 did the world find out. Laura was going to make a transcontinental trip—a new kind of transcontinental trip. However, months were to elapse before she got going. For one reason or another Orion either refused to take its place in the skies or, once it got there, slipped back to earth. Never did such bad luck follow any aviator as that which dogged Laura in her efforts to bring her princely plane back to the East. Dust storms held her up in the West, and when she did finally start from California she got lost in one of them. No wonder. That ferocious cloud of dust extended 22,000 feet above the earth! Later on she was forced down in Indianapolis by low oil pressure. And then, just as she made ready to start again, a firecracker exploded prematurely and injured her foot. The anti-tetanus serum which they were obliged to administer after that made her so ill that she had to linger in Indianapolis.

At last, however, one for the money, two for the show. On July 11, 1935, she achieved the long-awaited trip. At midnight Captain Ken Behr, manager of one of the East's busiest airports and official N.A.A. timer, sealed the barograph in her plane. Two hours later Laura herself appeared. The blue-green eyes under her beret sparkled with excitement. The slight figure in the aviator jacket was vibrant. First, she looked over every detail of the low-winged plane. Then she personally supervised loading it up with 600 gallons of gas and 40 gallons of oil.

The drama increased in the darkness of that midsummer night. Two fire trucks, a police emergency squad, and an ambulance had arrived on the scene. So had some personal friends, newshawks, and a throng interested in any flier about to depart for a record-breaking flight. Through the hum of conversation there was one constant note. Would the Orion ever be able to get up with that heavy load?

It was a natural wonder. When at 4:31 A.M. she finally got away it was with the heaviest load ever taken up by any woman. The 4,040 pounds' weight of the ship plus 640 gallons of gas and oil made it elephantine. And as the Orion moved off the spectators held their breaths. Yes, after a run of 2,500 feet in a cross wind it was lifting. Still, as it made a quick left-hand turn to double over the field and fly west —by Jove, the tail was dipping. Wasn't that mulish constellation, the Orion, going to take its place in the heavens, after all?

But, though Laura had a difficult assignment in getting altitude with her heavy weight, she did finally attain sufficient speed to level off. Keeping up a speed which never fell below 200 miles an hour and traveling for the most part on the radio beam of the T.W.A. route, she set down her black-cowled plane in California at 7:51 that night. And as the N.A.A. representative timed her in, there was every reason for the look of triumph in those aquamarine eyes. She was the first woman to span the American continent in a non-stop flight.

Even so, some bad luck in the way of strong head winds and electrical storms had held down her time to 18 hours, 20 minutes, and 30 seconds. That was below the record established by Amelia Earhart for a transcontinental flight —not nonstop—in 1932. And though she had bettered the time made by Collyer and Tucker in their Lockheed Vega in 1928 she was not entirely satisfied.

When she reached California it was buzzing with talk about prospective flights. Wiley Post was getting a passport for his hop to Siberia. Sir Charles Kingsford-Smith, who was expected to dock very soon, was reputedly planning an 11,330-mile flight from London to Sydney, Australia, in his magnificent transpacific monoplane, Lady Southern Cross. Various others had some project up their sleeves.

Into this atmosphere Laura Ingalls, who was never without a bee in her bonnet, fitted perfectly.

On July 19 she was one of a group composed of Amelia Earhart, Wiley Post, and Roscoe Turner. They were lunching together at the famous Union Terminal in Los Angeles and the air vibrated with their talk of gas consumption, weather, routes, and new types of planes. In the midst of this professional conversation Laura suddenly put forth her own pet plan. She wanted to break Amelia's record on her West-to-East flight.

"Why not?" asked Amelia in the most friendly fashion. "Aren't records made to be broken?" And later on, when Laura unfolded her plans for the project, she nodded her head approvingly. "That record has stood too long."

So Laura went along with her brand-new ambition. In order to fulfill it, she now got a new Lockheed Orion called *Auto-da-Fé*. And in the face of tradition on Friday, September 13, 1935, she started out from California by the dawn's early light and at 7:15 that evening she was saying to the officials at New York's Floyd Bennett Field, "Sorry to have kept you waiting so long."

Certainly she couldn't have been sorry about anything else. For what had she done? By crossing the continent in 13 hours, 34 minutes, and 5 seconds, she had beaten Amelia Earhart's elapsed flying time by nearly 3½ hours. What was more, she had missed smashing by only seven minutes and a fraction Frank Hawks's record of June 2, 1933. It must be remembered, however, that none of the three had eclipsed

Ruth Nichols' actual flying time. That brilliant record of 13 hours, 21 minutes was to stand until Jacqueline Cochran took to the skies with a faster and more nearly perfect plane.

Airwoman magazine of October 1935 contains an interview with Laura after she had made her record-breaking flight, and we quote from it directly:

Nothing happened. The only "high light" was a big full moon which rose right after I left St. Louis. I flew for hours in the darkness, and little Elmer deserves a lot of credit for the success of the flight. Records could not be broken if it were not for the advancement in design and equipment, high octane rating of fuel, which makes it possible to run an engine full throttle without burning itself up, and streamlining.

We hasten to explain that the "little Elmer" to which she refers in this modest statement is aviators' slang for the Sperry gyroscope. What, then, is this gyroscope? It is a descendant of the gyro-stabilizer which the brilliant young inventor Lawrence Sperry and his father, Elmer, introduced in 1914. But, whereas the older device merely kept the ship on an even keel, the modern gyroscope—or pilot—has become the "brain of the airplane." Thanks to it, the modern pilot may sit with folded hands and let George do it.

Uncanny, indeed, is its nervous system. Because of a pair of rapidly revolving gyroscopes—one vertical and the other horizontal—it senses any deviation from the set course and immediately rights it. Of course the human element is still

in supreme control. At any moment the automatism may
be shut off and hands carry on. This invention, which has
relieved so tremendously the strain upon the aviator and
which has also contributed to the smoothness of flying,
did not come into its own until 1935.

The next important news about Laura Ingalls comes in
1936 when she won the second prize—$2,500—in the famous
Bendix Race from which Louise Thaden and Blanche Noyes
emerged as victors. That she should have been able to enter
the race at all—much less, to capture a prize—this is testi-
mony to the almost fabulous endurance of this aviator. Late
in the afternoon of the day before the race she brought her
Orion into New York. She looked so spent on her arrival
that Ruth Nichols, one of the official starters of the contest,
insisted upon having a physician look her over. Oh, she was
all right. Only, of course, she must have rest. Such was the
doctor's verdict.

Rest? Laura had her own interpretation of that word.
There was a short interval when she lay quiet on the cot in
the hangar. Then she raced off in an automobile to Hart-
ford to buy some spark plugs for Orion. And when she
came back to the field—what time for a cot now? She put
in those new spark plugs herself. She went over every inch
of Orion's anatomy. It was after such hours of activity that
she started out for the grueling race across the continent.
In the entire history of sportsman aviation there is no such
record of endurance.

Blessed with great mechanical aptitude, she always insisted upon taking care of her own plane. In fact, she was fearful of any alien eyes prying into its mechanism. And about this night before the Bendix Race there is told a story in which many people find support for the theory that Laura Ingalls' political involvement was due to a real psychopathic maladjustment. Certainly the story demonstrates that she was without any emotional balance.

Among the reporters gathered on the field that night was one whose nose for news was somewhat overdeveloped. He drew near the cockpit where Laura was tinkering. "Keep away!" she roared. He paid no attention. He started to enter the forbidden zone. And at this she jerked from its holster the pistol which every aviator is licensed to carry. Needless to say, the reporter obeyed. Afterward he is quoted as saying, "Never shall I forget those blue-green eyes of hers. They blazed like a tiger's."

After the Bendix Race the rumors which she, above all other aviators, was always able to both light and explode began to multiply. She was going to set some new records for women which would keep the men shooting at them for years to come. She was going to fly here. She was going to fly there. Undoubtedly, too, she would have done some of these things had it not been for the aviator's chronic nightmare, "Who's going to finance me?" The only difference between her and the others who were handicapped in the same way was that they kept their dreams strictly private.

On September 26, 1939, she was again to make front-page

news. Alas! this time the aerial achievement was of a different character. She had bombed the Capitol at Washington with anti-war pamphlets. In her defense it must be said that few were the aviators who realized that there were certain restricted zones over which they could not fly. Nevertheless, she was obliged to appear before Roscoe Walter, C.A.A. trial examiner. Admitting that she had violated traffic regulations—albeit unwittingly—she maintained stoutly that love of her country's good was back of her action. Later she volunteered, "I didn't drop any monkey wrenches from the side of the plane. I've been brought up very carefully."

This scene merely foreshadowed the tragic events of 1941–42. Now so identified with the anti-war group that she was often called Isolationist Aviator 2, she chose a path for her fervor explicitly against the laws of her country. In February of 1942 she was convicted of being the unregistered representative of German agents from whom she was receiving pay. During the course of her trial it was said that she had been a constant and devoted reader of Hitler's *Mein Kampf*.

Did she honestly believe, as she so passionately asserted, that love of her own country was back of her offense? Did she really think that the happiness of America was dependent upon peace with Germany? Was she the victim of some hallucination that she was indeed a Mata-Hari, making use of her German employers to ferret out plots against her country? Some of her fellow aviators cling to the charitable

view, and they point to a hundred stories of her eccentrici-
ties to explain her confusion. In fact, they refuse to believe
that the last word has been said about the brilliant—although misguided—mind and the unquestioned mastery
of the air which is hers.

"She was a great flier." With these words of a famous
man pilot we shall leave the first woman to make a nonstop
flight across the continent, the first woman who ever sailed,
bold and free as the eagle, over the snowcapped, treacherous
Andes to win the greatest aviation honor either man or
woman can attain.

Anne Lindbergh

LISTEN! THE POET

☆

NINETEEN TWENTY-SEVEN. It was May in Northampton, Massachusetts. Tulips, gorgeous as Spanish grandees, lifted their silken red and yellow above the campus grass of Smith College. A talkative little brook glinted in the sunlight. And as Anne Morrow walked toward Northampton Meadows she was conscious of blooming dogwood, a spray of white against filmy green foliage.

Very often this slim, dark-haired collegian walked alone. Today, however, she was with a group of her classmates and, though the violet-blue eyes which were often mistaken for brown turned frequently from her companions to rest on sky and meadow, she was gayer than usual. When the group reached Northampton Meadows those eyes were suddenly alive with interest. One of two young men who had been tinkering with a somewhat decrepit plane hurried over to the bevy of Smith girls.

"How about it, girls? Fine day for a ride. Come along—take you up for a song."

The girls looked from the speaker, an operator of this flying field on the Holyoke-Northampton Highway to one another. Some grinned self-consciously. Others shook their heads. A few seemed to be considering the proposal. Among the last named was Anne Morrow.

"Aw, come on, girls," urged the salesman; "where's your nerve? Safe as the old rocking chair back home, and say, once you go up you'll never want to come down."

There was a moment's silence. Then Anne Morrow stepped forward. "All right," she said, "I'll go."

Shortly afterward an old Army plane was soaring into the heavens. And from its cockpit looked down on the Connecticut Valley dark blue eyes which wondered. Would she ever go up again? So may Anne Morrow have asked herself on her first flight.

1927. It was May in St. Louis. Already the summer heat which blankets the old frontier town was beginning to hang over the river. But the hint of sultriness did not rob a certain young man either of his springing stride or the light of stern determination in his eyes. Charles A. Lindbergh was making final preparations for his flight across the Atlantic. It was to be no ordinary hop. No mere crossing between Newfoundland and Ireland would do. Although this valorous feat had won just acclaim for the young aviators who had performed it, a certain American citizen of French birth was eager for something more. Mr. Raymond Orteig, owner of the Lafayette Hotel, for years a haunt of gourmets, intellectuals, and all those who like to meet their France in

America, was now offering $25,000 for the first nonstop flight between New York and Paris.

Twenty-five thousand dollars! To Charles A. Lindbergh, a young mail pilot, the sum was dazzling, for he had been brought up in a modest home in Michigan. What was more, he heard in that modest home no admiration for wealth. His father, Swedish-born representative for Congress, was forever protesting against the standard, "How many millions are you worth?" He would have liked to replace it with, "How much good have you done for mankind?" Brought up in this atmosphere, it is doubtful whether Lindbergh thought of $25,000 as an end. But—he was a flier. That was his destiny. He had left college after his second year because he could see no other future. And to be a flier—unless you were content with some little salaried job—you had to have money. A good plane. Some influence. A record. No wonder this young man of twenty-five strained toward that $25,000 offered by Raymond Orteig, the French-born citizen of New York. His whole future was at stake.

But what right had he to hope? A good flier? Yes, he was that, and he knew it. But there were dozens and even hundreds who had as excellent a record as he. And opposed to him in this race? Admiral Byrd, backed by wealth and influence, was making ready to fly eastward from Long Island. He and various others were pitted against Lindbergh for the $25,000 prize.

However, there were people in St. Louis who believed in the flying genius of this unknown young man. It was with their help that he set off on that night of May 20, 1927, from Mitchel Field near New York City. His departure was unsung. Metropolitan newspapers knew nothing of Charles A. Lindbergh of St. Louis and cared less. The very next evening they realized their blindness, for the young mail pilot who had left New York City was the hero who had landed at Le Bourget Field just outside of Paris. He had made the first nonstop flight between the metropolis of America and the metropolis of France.

The May of Northampton and the May of St. Louis— these were cross-sections of our American scene. Between the Smith College girl taking her first ride in the air and the mail pilot who won fame and fortune in a single day there is no obvious connection. But a playwright called Life decided otherwise. To weave the two Mays into a whole, to make two widely separate human beings into a pair which will always conjure the imagination—this was the finished masterpiece of Life.

Let us, however, go back to the Smith College girl of that May day. She was born in Englewood, New Jersey, on June 22, 1907. Although her father, the late Dwight W. Morrow, had the reputation of being a wealthy man—was he not at one time a member of the firm of J. P. Morgan and Company?—she was brought up almost as unpretentiously as Lindbergh himself. Save for one year at Miss Chapin's in New York City, she had attended only local schools before going to Smith College. And when she "came out" there were no thousands invested at Pierre's or Sherry's. The modest home in Englewood—that was the scene of Anne's debut.

As to the choice of Smith, this was no accident. Her grandmother had started the tradition. Then came her mother, the vigorous-minded woman who was later to be acting president of the great Massachusetts college. And before Anne entered, her older sister Elizabeth had paved the way. It was natural, therefore, that she should feel very much at home in the classrooms at Northampton.

But it was not the collegiate "good times" which ever attracted her. That week-end "date" which is the meter of a modern girl's popularity—this meant nothing to Anne Morrow. She had, in fact, little to contribute to those parties where collegians of both sexes meet on a basis of "School's out. Now let's go to it." She did not smoke nor drink. She cared little for dancing. And as for those nice boys from Harvard and Yale and Williams about whom other girls raved, these must have seemed far less attractive than Tancred or Roland the Warrior or the other heroes she met in books.

For the most accurate portrait of Anne Morrow as she was in her Smith days we had perhaps best turn to her own pen. In the college magazine of October 1926 there appeared "Caprice," verses which she wrote after seeing Raquel Meller. That she saw herself as a Quaker—this is beyond doubt. However, the view is so tinged with gay humor that we can never blame her for self-satisfaction. Indeed, these verses show you what fun brimmed always in the heart of Dwight Morrow's daughter. No, it was not the kind which makes a girl the life of the party. It was the whimsical and rarefied fun of an earlier American woman poet—Emily Dickinson. Here are the lines:

I should like to be a dancer,
A slim, persuasive dancer,
A Scarlet Spanish dancer,
　　If you please.
But he said, "Just now we're crowded—
With these Carmens—simply crowded—
I can't find"—his forehead clouded—
　　"Vacancies."

"I suppose you want to tango—"
And he sighed—"or a fandango,
Scarlet cigarette and tango.
　　Scarlet smile.
In a century or twenty,
We may want you. We have plenty
　　For a while."

There's a place for Quaker maidens
For brown-eyed Quaker maidens
For blue-eyed Quaker maidens
　　There's a place.
So I play the role of Quaker
And I do not blame my Maker
For I think I wear the Quaker
　　With a grace.

With the quotation of this verse we reach the very kernel of Anne Morrow's being. She is and always has been a poet. At Smith her gift was acknowledged by frequent publication in the college magazine. Indeed, during this year of 1927, when we first meet her, there was hardly an issue which could not boast some contribution of hers. She was then a junior, and people were already saying, "I bet Anne gets the Mary Augusta Jordan prize next year." She did win it—that coveted honor awarded annually by the alumnae to the senior who at some period of her undergraduate work has done the most distinguished piece of literary work in either poetry or prose. As a matter of fact, her prose always had the same exquisite feeling for words as had her verse. And another prize which she won was for her essay, "Women of Dr. Johnson's Time."

However, she was not entirely absorbed by the pen. She always had a keen interest in current events and was an active member of the French Club, and she had no trouble, of course, in making Alpha, one of the two honor societies at Northampton.

Too bookish a young person? In fact, almost a prig? Not a bit of it. Although she was very shy, although she always found it hard to give herself to the usual person, those who knew her well were her devoted friends. "Anne always took such a terrible photograph," one of these friends said not long ago as she leafed over the yearbook of 1928. "You can never get any idea of what a clear, lovely skin she had and how beautiful her eyes were—sometimes very gay, often very sad, and always—oh, so intelligent."

For her Christmas vacation in the year of 1927 she went down to Mexico. There her father had become the American Ambassador and the family was flavoring with delight the color and charm of our neighboring Southern land. Anne had in mind perhaps some fresh explorations in this beautiful Mexico City when her father received news. That news was quite enough to banish sombreros and splendid golden altars and balconies of grilled iron from the mind of any college junior. Charles A. Lindbergh was now making a good-will tour of Mexico and Central America. He was going to call on the Morrow family.

Lindbergh, the hero of America! If shy Anne Morrow had been asked to step up to the stage of the Metropolitan Opera House and trill a response to triumphant Radamès surging forward in his chariot to the beat of *Aïda's* great march, she probably would not have felt more overwhelmed. "Ah well," she may have thought, "I won't have to talk. Elizabeth will do that." For socially she was always more quiet than her elder sister.

What did she think of him at first meeting? What did he think of her? Was he able to break down her shyness? Did she ever get to the point where she could tell him about her one-and-only ride in the air? These are questions which cannot be answered. We know only the epilogue of that meeting.

Anne was graduated from Smith the following year. Yes, of course she wrote the "Ivy Song." But meanwhile her verses had begun to appear beyond the college campus. And in April 1928 *Scribner's* magazine published the following:

When I was young I felt so small
And frightened, for the world was tall.

And even grasses seemed to me
A forest of immensity

Until I learned that I could grow,
A glance would leave them far below.

Spanning a tree's height with my eye
Suddenly I seemed as high.

And fixing on a star I grew—
I pushed my head against the blue.

Still, like a singing lark, I find
Rapture to leave the grass behind.

And sometimes, standing in a crowd,
My lips are cool against a cloud.

Anyone familiar with the habits of magazines does not ask whether this was written after her meeting with Lindbergh. Almost certainly it was not. For publication hardly ever follows immediately upon acceptance. Besides, this poem does not speak of a girl who has been converted to flying. That one doesn't need mere mechanical wings to soar, that thought may lift one above the trees to the clouds—this is the message of "Height."

Yet before her graduation Anne had gone up with Lindbergh—not once, but frequently. For, having visited the Morrows at the Christmas season of 1927, the victor of the air came back—again and again. During the summer of 1928 the fact that Colonel Lindbergh was in Mexico City no longer rated as news. During that summer people began to tamper with Shakespeare and to say, "To Morrow and to Morrow and to Morrow leads—why, of course, to Matrimony."

But the Walter Winchell minds were puzzled. Which was it—Anne or the attractive elder sister who had preceded her at Smith? Perhaps Anne herself may have been in doubt. Indeed, there is a story to the effect that when Lindbergh asked her to marry him she said, "But I always thought it was Elizabeth." If this be true, it is in line with the humility with which this gifted girl always met praise. Why should America's hero choose her—the "Quaker maid"—when "slim, persuasive dancers" were throwing garlands at his feet?

And didn't this girl who had always lived in her bower of books find something lacking in her hero? For certainly Lindbergh cared nothing for the authors who had meant life itself to her. He is and always has been a factual person whose one-track mind has brought him success. He could no more meet Anne Morrow in a discussion about

Keats and Shelley and John Donne than he could follow
the paths of her elfin humor. Hasn't he always been no-
torious for practical jokes?

It may have been that she was more dazzled because of
the very gap between them. What had she done but read
about heroes? Here was the hero himself, theme of song and
story for all the scribblers who followed him. The old
fascination which the man of action exerts upon the
dreamer was never illustrated more eloquently than in this
case.

They were married in the early summer of 1929. Just before this they were victims of a minor accident. Flying together one day, Lindbergh discovered that they had lost a
wheel.

"When we land, we'll overturn—but don't be afraid,"
he said quietly. Then with equal calmness he placed her
in the back seat and tucked her in among cushions which
he hoped would break the fall.

Sitting there in the back seat, Anne stared at the broad
shoulders and the head with its wind-blown hair. She knew
what he was trying to do. It was to land on one wheel and
the tail skid. She knew, too, how dangerous was that task.
But how could she be terribly frightened? What power
there was in every look and every movement of this man!
Siegfried the fearless! Again there must have swept over
her, more intoxicatingly than ever before, the sense of his
perfect mastery.

The plane did overturn. It was, in fact, badly damaged.

But, though Lindbergh had been hurled against the side of the ship, he managed to crawl out from the wreckage and help Anne through a window.

Many people think of Anne Lindbergh as merely the accompanist for a master musician of the air. Nothing could be farther from the truth. True, she acted as copilot, navigator, and radio operator on all their historic *Wanderjahre*. But in her own name she is also an expert pilot.

She made her first solo flight August 24, 1929. Her husband had been instructing her ever since their marriage a few months earlier, and on the morning of this particular day he stayed in the air with her for an hour. Then when the ship—a Curtiss Fledgling—came down for one of its practice landings he clambered out. It was to stay out. After a few final instructions he waved to her and she went aloft —for the first time alone. Exultation. But in it there was no fear. Circling the field of the Aviation Country Club in Hicksville, Long Island, she was conscious always that *he* was below. What could happen when he was there? She landed to be sent aloft by him again. But this time when she came down he was walking away. She was to fly in solitude. A child leaving the adult finger which has always guided its uncertain steps, she flew into the ether.

Undoubtedly her husband's confidence in her helped still any doubt as she rose. He had, indeed, never a fear for his pupil. All the time Anne was making her first rapturous solo flight he sat reading a newspaper on the clubhouse veranda. He was just as sure of her the following January when she sat alone in a motorless plane perched on the top of a California hill.

A few minutes of tense silence. Then came a signal. The elastic cords which held the plane were released and out into space soared beautiful birdlike wings. Beyond sparkled the blue Pacific, and as Anne glided toward it . . . "Alone upon a peak in Darien" . . . Did old, well-loved words of Keats sing in her heart? Was she tempted to let herself drift on and on over Balboa's ocean in this world of silence unbroken by any hum of machinery? If so, she did not yield to the temptation. Three minutes. Then she wheeled to the right in a 90-degree bank. Then to the left in an equally steep angle. At last a glorious swoop brought her down over a meadow. A perfect landing.

She was wearing, that day, an old pair of overalls big enough for a six-foot man, and as she got out of the glider she looked like a happy urchin. Lindbergh, who had been watching her tryout in this sail plane—the same in which Hawley Bowlus, its builder, had recently established an American gliding record of six hours, 19 minutes, and 3 seconds, and in which he himself had just won a first-class glider pilot's license—beamed back at her. True birds of a feather, they were ready now to fly together to farthest land and sea.

As a matter of fact, Anne's flight of six minutes and one tenth seconds qualified her for a first-class glider pilot's license. She received it and thus became the first woman in the United States to win the title, one which few men could then boast. Most people do not realize this. Neither are they aware that on May 29, 1930, she was given the highest grade flying license possible to obtain.

This is the one called transport, and it was awarded to her after she had completed precision landings, spirals, and other trick stuff of flying. Most perfect of all her maneuvers were the figure eights. Flying at 1,500 feet, she turned in and out between the live wire and gas tank which served as her pylons with unexcelled accuracy. At the end of that performance, which lasted for fifteen minutes, it was no wonder that the inspector remarked, "A flight test well flown."

Such mastery of the air involved patient practice. Yes, but it was rewarded by a rapture which she herself expresses so passionately in her book, *North to the Orient:*

... the engine roared on again ... up, up, up ... I felt myself gasping to get up, like a drowning man. There—the sky was blue above—the sky and the sun! Courage flowed back in my veins, a warm, pounding stream. Thank God, there is the sky. Hold on to it with both hands. Let it pull you up. Oh, let us stay here, I thought, up in this clear bright world of reality, where we can see the sky and feel the sun. Let's never go down.

Far different is the cry sounded by this veteran of the air from that of the college girl who had written that those who stand upon the ground can "cool their lips against a cloud." After being the literal comrade of clouds and peaks the spirit's flight now seemed perhaps a trifle pallid.

Even more arduous than her work at the wheel was her study of navigation and radio. Never did the girl at Smith bone so hard on any course as did the woman of the air on her technique of radio receiving. Later on she was to describe her difficulties in the words:

. . . the dots and dashes refused to stay on my pencil mark. I found I needed one hand constantly on the main dial, another on the vernier, trying to pin down my station like an elusive butterfly. I wanted a third to write the message, and still another to hold the pad—the work of four hands to be done by two incompetent ones. That meant acrobatics.

However, things came easier after a while and we find her writing ecstatically of the time when she began to hear something else besides cosmic crashes. This is how she describes a great thrill:

. . . faint squeaks against the welter of noise, precise scratchings upon the blurred surface of sound. So dim and faint, they were no more than a twig's tapping on a window pane during a storm; no more than a crab's track on sand, partly erased by a wave; or a dead leaf's tracing on new-fallen snow. They were living, however; they were human, I was sure. They were dot-dash, Morse-code letters, words, messages of a human being.

I put my pencil on the page and let it jot down stray letters where I thought I could hear them between crashes: "O—T—C—FN—R—K—L." Then suddenly, through the welter of sounds, I heard no longer letters but my name, or so it sounded to me: KHCAL, the call letters of the plane. Across an ocean and through the night, my name! More thrilling than to hear your nickname in a roomful of strangers, or your own language in a street of foreigners. . . .

Someone had heard us—in our little cabin, flying in the dark in a plane thousands of miles away. . . . "Dit darr dit—dit darr dit." . . . We were really in contact with South America. We had jumped the distance, touched hands between hemispheres.

Although Anne acted as copilot in the flights which gave supreme fame to this flying pair, her greatest contribution was the skill in navigation and radio technique which she had so laboriously acquired. In the team's first record this skill may not have been accented to any great degree. For that was on Easter Day, 1930, when they broke Frank Hawks's record by crossing the continent in 14 hours, 45 minutes, and 22 seconds. Soon afterward, however, Anne's place was to become more clear-cut.

Lindbergh was now an advisory of the Pan-American Airways, and a survey of air-mail routes led him far down Central and South America. He took Anne with him, and together they flew seven thousand miles. It was then that their minds first took fire from the remains of ancient civilizations. They were especially fascinated by the culture of the Mayas, and when they launched a second expedition over the treacherous jungles of Guatemala they made a great historical find. It was the Temple of the Warriors. Anne took photographs of this, as well as of many other archaeological treasures. These, as well as the detailed information which they were able to supply, proved of the greatest help to subsequent research.

By these two flights the Lindberghs soared into more stately realms of aviation history. Here was no mere flying for the sake of a record—speed, altitude, endurance. Here was flying with a real historical and geographical meaning. And it was in the same spirit that they approached one of the most celebrated of their journeys, the one which Anne was later to embody in the first of her beautiful works of prose. They started it early in the summer of 1931.

Taking off from College Point, Long Island, in their sea-plane, they flew by way of North Haven, Maine, and Ottawa to a tiny fur-trading post at Baker Lake, a spot where no white woman had ever ventured before. Then, skimming along the coast of Canada, they came by way of Aklavik to Point Barrow, that isolated settlement waiting for the ice-blocked boat to bring it its year's supply of provisions. Next Japan. And last—river-covered rice fields, millions of starving people—China, patient, timeless China in the throes of one of her most terrible floods.

In China Anne gave up her seat in the plane. It was to make way for a Chinese and an American doctor with medicine for the sufferers. After bringing these passengers up the Yangtze River, Lindbergh landed just outside the city wall of Nanking. Beside it numerous sampans were floating, and into one of them climbed the Chinese doctor. A sack of antitoxin from the plane was pushed in after him.

But those hundreds of desperate starvelings misunderstood. "Food!" The magic word passed from boat to boat. In vain the Chinese physician tried to make them understand he brought only medicine. They swarmed over his boat and sank it. And as he came back to the plane where Lindbergh and the American physician were trying to hold a mob at bay the scene became violent. Men were leaping from boat to boat, toppling over each other in an effort to get nearer that machine which they still imagined must carry food. Some who really reached the plane clung to wings and tail surface.

"We're going to start the engine!" shouted the Chinese doctor. "If you don't get back you'll be killed!"

Even that warning failed to budge them at first. Only when they heard the roar of the motor did they begin to fall back. They had asked for bread and been given—antitoxin.

Although Anne did not witness this scene she conjured it vividly before her. In fact, she was lacerated every moment of her flight over China by this vast human misery. Her poet's heart made her one with each hunger-racked creature whom she saw, and she thought how desperate those people must have felt when they saw their last hope disappear into the clouds above them. Or how those hundreds of others felt as they saw a plane skim over their isolated country. It tore at her heart to think of their helplessness.

This journey was to be cut short by the death of Anne's father. Upon their return Lindbergh continued in his ad-

ANNE MORROW LINDBERGH, internationally recognized flier and writer.

ANNE MORROW LINDBERGH with her husband, Charles A. Lindbergh, on one of their famous survey flights.

visory capacity to the Pan-American Airways. As for
Anne, she was busy with her pen. *North to the Orient*—that
was the title of the book in which she described their
recent wanderings. No sooner was it published than she
was hailed as one of the real masters of English prose who
have appeared in our century. What was more, here was
the first master to celebrate the poetry which lies beneath
that man-made mechanism, the plane. *North to the Orient*
was destined to become more than a best seller. It is a
classic.

In the meanwhile the Lindberghs had had their first
child. He was a Viking son with blue eyes and golden hair
and they had just moved with him into their New Jersey
country place when the tragedy occurred which shook the
whole world and perhaps changed the course of the Lindberghs' destiny. The beautiful boy was kidnaped and
many weeks later his body was found. Agony of suspense
followed by the cold steel of certainty. Weeks when tragedy
was stalked by gaping crowds. A trial of the murderer
which dragged on in an atmosphere of horrible showman-
ship. Such a situation would have eaten into the hearts of
any other two people on earth. But in the case of the
Lindberghs torture was supreme. For, in spite of many
temperamental differences, they had this in common: they
both gave themselves to few people. They both felt loneliest
in a crowd. To them both the publicity of their loss was the
crux of agony.

In these days intimate friends tell us that "Charles" was

the only person on earth who could comfort his wife. Their crucifixion drew them closer together, in fact, than ever before. Perhaps, too, it made them feel more alien to the multitudes than ever before. At all events, together they rose to that spiritual height which says, "I am the master of my fate; I am the captain of my soul." Soon after a second son was born to them they launched upon one of the most celebrated of their voyages.

It was in July 1933 that they began that survey of North Atlantic sea routes which took them first from Labrador to Copenhagen. Subsequently they flew along the coasts of Norway, Scotland, Ireland, Spain, and Portugal. From this last-named country their course was charted south to Africa. Over the equator then to South America and from South America home. A comprehensive route and one destined to yield more than mere commercial harvest.

During this trip people who met Anne Lindbergh in Europe had always one thing to say: "She has the saddest eyes in the world"—that was the immediate impression made before even her voice, so soft and well modulated, could register. But, though she might be marked deep by the loss of her first-born, the fine heroic temper of her soul was not broken. Far from it. She was to create from the flight over Africa and South America a book even finer than her *North to the Orient*.

In this second masterpiece, *Listen! The Wind,* she introduced many colorful details. There she was pumping out fuel from an overloaded plane. There were both of them creeping out of houses at dawn. But we forget these in the cosmic magnificence of her descriptions. Who, for example, will ever forget the time when they had waited so long for a wind to help them? At last a little breeze came in the night and they started away in its black darkness. At first their engine sputtered. And then . . . Let Anne Lindbergh tell you the rest in her own prose:

. . . it smooths out now, like a long sigh, like a person breathing easily, freely. Like someone singing ecstatically, climbing, soaring—sustained note of power and joy. We turn from the lights of the city; we pivot on a dark wing; we roar over the earth. The plane seems exultant now, even arrogant. We did it, we did it! We're up, above you. We were dependent on you just now, River, prisoners fawning on you for favors, for wind and light. But now, we are free. We are up; we are off. We can toss you aside, you there, way below us, a few lights in the great dark silent world that is ours—for we are above it.

Anne Lindbergh has been honored by many grave assemblages. She was one of the few women to whom the Cross of Honor was awarded. "As radio operator and navigator and at times relieving the pilot—as bringing the nations visited into friendlier contact, she helped establish bonds of understanding and sympathy, to promote the closer unity of Mankind"—these were the words which accompanied this citation. Even more comprehensive and flattering was the tribute paid her when she was given the Hubbard Medal of the National Geographic Society.

By winning this medal, which had been awarded to men like Admiral Byrd, Admiral Peary, Sir Ernest H. Shackleton, Captain Roald Amundsen, and her own husband, she became the first woman in the world ever to be so honored. And the words on the medal sum up tersely why she should have been so distinguished. "In recognition of your courageous and skillful work as copilot and radio operator during 40,000 miles of flight over five continents and many seas."

However, most of us like better the definition of her gifts made by her own college when it bestowed upon her an honorary degree. "She gave wings to poetry"—that is what Smith said to the woman who once, as a shy, dreaming girl, had written from her bower of books. For she, like the French writer, Saint Exupéry, has been able to communicate to the world the rapturous adventure of the sky. What is more, every word she writes mirrors the courage and loyalty and the tenderness for all mankind of a soul whose integrity can never be questioned.

Jacqueline Cochran

A DYNAMO OF DREAMS

☆

The place is Cleveland. The year is 1938. The month is September. The occasion is the National Air Races and the finish of the feature event.

A small army of men, wearing badges and ribbons marked "Judge" or "Official," bustle about a silver pursuit monoplane and the young woman standing beside it. Couriers on motorcycles whisk across the airport to the official timer's stand and back to the plane. Photographers cry, "This way, please!" "Smile, please!" "Once more, please!" The crowd is roaring. Overhead a formation of Army planes dive and swoop in salute. And a myriad of aviation editors and feature writers from all over the nation press forward, clamoring for a statement.

Turning from the excited turmoil about her, the young woman eagerly scans the crowd. "Where's my husband?" she asks.

At the question a signal is given, a car dashes up to the grandstand boxes, pauses an instant, and speeds across the fields to stop in front of the racer. From the car leaps a bespectacled man who looks like a professor. "Jackie!" he cries, and gathers her close. Her arms go about his shoulders.

With a cry of "Wow!" the photographers snap their shutters. One reporter asks another, "Who's the lucky guy?"

For a moment nobody identifies the man who has received such special attention from the race committee and the warm greeting from the winner. Yes, you may have guessed it. The romantic-looking young woman has just won the Bendix Race.

On any other day this man could command the attention of any committee in the country. He could have bought outright the flying field and all the planes flocking around it. For he is the "boy wizard" of Wall Street, the man who transformed an investment of $39,000 into a fortune of more than a million during the depression. His name is Floyd Odlum, and he is president of the Atlas Corporation.

On race day, however, not even a millionaire could violate the sacred rules of the airfield. Only at the request of the small blonde-haired pilot could he leave the spectators' box for the field. But at that moment her word amounted to command to airmen and officials alike. The only woman competing this year, she has crossed the line twenty-three minutes ahead of her nearest competitor and won one of the most difficult long-distance races in aviation annals. Not everyone knows that in private life she is Mrs. Floyd

B. Odlum. This skilled pilot and most-publicized woman flier goes by the name she herself made famous, Jacqueline Cochran.

Standing with her back to the crowd, Jacqueline retreats for a moment into that personal self. Happily she received her husband's ardent congratulations. She has only a moment, for her task is not finished. She must be up and off to establish the women's transcontinental record and take the additional prize of $1,000 offered to the first contestant in the race to reach Bendix, New Jersey.

"Go it, Jackie," urges her husband with a smile of pride. He helps her into the little plane. "Take some tomato juice. Good girl! I know you'll win. Not a racer in sight yet."

With a whir of propeller and roar of engine, the silver monoplane tears down the field and rises skyward. That night Jacqueline Cochran's name was on the front pages of every American journal. She had whisked over Bendix and landed at Floyd Bennett Field with an official record time which cut three and a half hours from the best previous flying time made by a woman. Beating ten of the world's best racing fliers, she had met this supreme test of pilots, planes, and parts with a record of 10 hours, 7 minutes, and 10 seconds.

Thousands of people who read those headlines said to themselves, "Who is this amazing young woman anyway?"

The answer reads like a scenario. Born near Pensacola, Florida, about thirty years ago, she was orphaned at an early age. Sheltered by a moneyless but friendly couple, she helped with the housework in return for board, lodging, and pin money. At fourteen she was a real wage earner with a job in a beauty shop which paid $35 a week. After a few years she had her own shop, but, although it was a success, Jacqueline decided to take nurses' training, and at twenty we find her a registered nurse. That was not her destiny, however. She knew it and came North to conquer new fields. After a period of special coaching in Philadelphia, she secured a position in New York with the famous French *coiffeur,* Antoine. For a few years she worked with him both in New York and in Miami. But, nothing if not modern in spirit, Jacqueline was becoming air-minded. Watching the come and go of planes in Miami, reading about the achievement of Ruth Nichols, Amelia Earhart, and a score of others, the girl made up her mind that somehow she was going to be a flier. This idea was warmly supported by her friend Mabel Wildebrandt, the lawyer, who took a great interest in the gifted girl. Why couldn't she find a way to combine business and this new ambition? At last, with a beauty shop in the South bringing her in an income, she took her savings and went to New York with an idea for promoting cosmetics.

"If I learn to fly at my own expense and go on a country-wide tour to advertise your products, will you help support the expenses of my plane?"

That was the question Miss Cochran asked manufacturers in New York. Interest in her plan was sufficient to induce her to go ahead. And at that psychological moment came the wager. She made it at a party where for the first time she met Floyd B. Odlum. Talk turned on flying, and the brown-eyed girl with wind-blown hair said, "I'm going to start learning to fly and I think I can learn in three weeks' time."

"Is that a bet?" Mr. Odlum's keen eyes looked at her through his horn-rimmed glasses with amused admiration. "If so, I'll take you up on it."

Jacqueline Cochran flashed him the smile which nowadays beams at you from so many photographs. She, who by personality, grit, and cleverness had won her way to a margin of economic independence, felt she was betting on a certainty.

So she was. Next morning she was enrolled in a flying school. Two and a half weeks later she passed the Government test for a pilot's license. Then out to California by automobile for a winter course in a flying school. To get more individual practice she bought a battered old plane and when, through motor failure in a take-off it crashed into a fence, she acquired a brand-new Waco. It was delivered to her in the East where she had returned for further instruction.

Now began the hard and tedious work and the valiant effort necessary to get her transport license. Once through that stiff examination, Jacqueline thought she could fly. But on a trip West she got lost over the mountains, ran out of gas, and had to set her plane down on a green strip in the heart of the Rockies with damaged wing and landing gear. As soon as the plane was repaired Jackie, who must take a dare even from a mountain range, flew it over the Rockies again. The experience taught her how much more training she needed in instrument flying and navigation. Engaging as instructor a skilled pilot from the T.W.A. Line, she took intensive coaching for more than three months.

Such were the milestones on the path leading to 1934 and to Jacqueline's determination to do something big in aviation that year. Opportunity offered in the international race from London to Melbourne, Australia. Her instructor agreed to act as copilot and navigator and joyously she bought a Northrop plane with special turbo supercharged equipment. Then came the first check in this ill-fated enterprise. While it was being flown East for shipment to London, the plane crashed.

Was Jackie daunted by this ghastly mishap? Not she. Another plane was hastily chosen. True, it wasn't quite completed. No matter. Mechanics could finish construction on the voyage across the sea. There was less than no time to spare. Only two hours could the luckless pilots get to try out their plane before the race began. Small wonder, therefore, that trouble developed before long. In spite of a start well up among the best competitors, the two venturesome Americans had to land their plane near Bucharest—a descent made dangerous by failure of the flaps to work properly.

Returning from interviewing the mechanics about the

necessary repairs, the copilot looked at his companion with despairing eyes. "No use, Jackie," he said; "we can't make this race. We're out."

This costly disappointment was only one test of the young flier's will to aviation accomplishment. In 1935 she persuaded the entrants of the Bendix Race to let her compete. She might have succeeded, but motor trouble brought her down in Arizona. Next year, before she could enter the race, her plane caught fire during a practice period over the airport in Indianapolis. Only by cool judgment was she able to land safely. Then she helped attendants extinguish the flames.

Nineteen thirty-seven, however, was a different story. Then all her experience began to tell. Moreover, she had aroused the interest of Major Alexander de Seversky, the great genius of the air, who had for several years been developing planes in this country. He arranged to have Miss Cochran use one of his fast Army pursuit planes with a double-row Wasp engine. He wanted her to establish a new speed record for women. But before that attempt the young woman again entered the Bendix Race and this time won a third place. The success was all the sweeter because she had piloted the plane for five hours through the blackness of night. About a month later she established a new world's record in the Seversky plane at an average speed of 293 miles an hour. Next, in the same plane, she skimmed from New York to Miami in 4 hours and 12 minutes. At a speed of nearly 300 miles an hour she beat the record set by the best male pilots. Finally, she topped off the banner year at the close of the Miami Races by setting a new speed record for women at 255 miles an hour over a 100-kilometer course.

Outstanding woman flier for 1937. That's what she was. And on highly competent authority. For early in 1938 the girl who had taken her first lesson six years before was awarded the Clifford Burke Harmon Trophy by a committee of the International League of Aviators. Every year the Harmon awards are given to the most eminent men and women fliers. Then, indeed, newspapers, newsreels, and magazines did their all to familiarize the American public with this new star of aviation. Her arresting dark eyes, her dazzling smile, her slim figure—effective alike in slacks or smart frocks—made her decidedly pictorial. Real achievement made her a news feature.

As the camera recorded her taking from the hands of Mrs. Franklin D. Roosevelt the beautiful Harmon Trophy, Jacqueline looked like a delighted little girl. News accounts, however, told another story. The beaming young person was not only an outstanding flier, but head of a successful cosmetic business and the wife of the man who had once dared her to learn to fly in three weeks. As Mrs. Floyd Odlum, she presides over a New York City apartment high above the East River, a date ranch in California, a sheep ranch in Arizona, and a residence in Washington. In short, here was a modern self-made Cinderella. She had acted as her own fairy godmother. Out of a cosmetics jar she had fashioned a plane, had dashed off in it to the aerial ball, and had won the heart of the millionaire prince of Wall Street.

Where most legends end, however, Jacqueline's tale went right on, for in 1938 she won the great Bendix Race over all comers. The next year we find her setting a new national record, breaking Ruth Nichols' record by reaching an altitude of 30,050 feet. Then one after the other she set four new speed records, each in a different type of plane. The climax of the year was making the first blind landing by a woman flier. It was inevitable that once again she received the Harmon Trophy from the International League.

Catching her on the wing between her cosmetic factory in New Jersey and a hop to California, a newspaper columnist said to her, "Miss Cochran, why do you stay in business? You certainly don't need to do so, and it keeps you always on the go. The wonder to me is that you stay so well and look so fresh and young."

Jackie laughed. "Well," she explained, "about that fresh look—you see, I always sleep a lot and exercise before a flight. Afterward I take a beauty treatment. It rests me. What's more, I often take a draft of oxygen. I have a tank in the apartment as well as in my plane. It's the best possible pickup.

"As for staying in business," she went on, "I like it. It interests me. And then Floyd bet me I couldn't make a go of my cosmetics factory and I just wanted to prove I could."

Her packaging, her kit bag, the names she gives her products, all show originality. Stores like to feature her cosmetics because of her promotion work. She has flown 70,000 miles in a single year to cast her flier's prestige about the wares on display. Into San Antonio she flies, heralded like a princess royal and greeted by cheering crowds. The newspapers write long stories about her, adorned with pictures and flecked with admiring comments. Buyers treat her with respect. Customers flock in—hoping that perhaps a Cochran lipstick will by magic transform them into either beauties or fliers. Leaving a town full of admirers, Jackie whisks on to Minneapolis and repeats the whirlwind sales campaign.

Right here it must be said that unlike the typical success story, the real portrait of Jacqueline has surfaces deeply hidden from the public eye. Her many imaginative generosities to fliers in trouble are known only to the few. Furthermore, she has a secret ambition very different from those stressed in the usual interview. This is to found an orphan asylum some day where children can receive the individual care and affection she herself was denied as a little girl. For this creative enterprise Jackie hopes to earn the money all herself. Such unsuspected impulses in this young woman of brilliant action are like the shy colors lurking within the clear facets of a diamond.

Yet never does any other interest block that serious business of establishing new records. In 1940 she beat the international speed record for 2,000 kilometers by flying

JACQUELINE COCHRAN, twice winner of the Harmon Trophy as the outstanding woman flier of the year.

She is shown in her uniform as a leader of the American women pilots serving with the British Transport Auxiliary to the R.A.F.

at 332 miles per hour. This was after bettering her own 100-kilometer record for women. Veterans of the World War gave her a party to celebrate the event and Mayor LaGuardia presented her with a trophy from the Air Service Post of the American Legion. That tribute was the prelude to the unique honor given her a few months later. For the third time she was awarded the Clifford Burke Harmon Trophy. Never had that happened to anyone in the world before.

But Jacqueline Cochran was not content with piling up the scores. Not at a time when the most tragic events of all history were tearing the world to bits. Her mind wrestled with plans to find a way for American women fliers to enter the picture of preparedness and of service in the war against Fascism. First of all, she felt she must see with her own eyes what the English women fliers were doing to help their country. Interviews with British representatives here encouraged her to propose a daring plan for reaching London. Finally it was accepted. In June 1941 she flew a Lockheed Hudson bomber plane from Canada to England. Not alone, of course. With a radio operator and Captain Grafton Carlisle as navigator, the slender Jackie with her windblown hair sat at the controls all the way across the Atlantic.

That's how she got to England. Reporters flocked around her. But she was able to persuade the courteous photographers not to take her picture in her rumpled pilot's slacks. Bathed, and smart in a redingote print, she gave the desired interviews. Then she set out to see London in wartime. First she had to say a word to Lord Beaverbrook. Through her husband she had already exchanged dinners with him and he greeted her with a cordial, "How d'ye do, Cochran. Tell me everything that's going on in America."

This dynamo of British war production wanted her to pay a visit to his country place. But she chose to accept another and more thrilling invitation. This was to spend the day at a fighting station as guest of the R.A.F. Jacqueline's real business, however, was to observe what English women fliers were doing, and she was shown every type of activity undertaken by members of their two leading organizations, the Women's Auxiliary Air Force and the Air Transport Auxiliary. Women in the latter group were flying planes from factory to fighting stations. The WAAFs worked at coding, folding parachutes, clerical duties, and served as fire wardens and even as mechanics. In the repair of delicate instruments the skilled hands of women and their patience proved particularly valuable.

Pauline Gower, leader of fifty women pilots, told Jacqueline she needed far more help than she could get. Interviews with men and women engaged in tasks of great responsibility, a bird's-eye view of bombed London, and a comprehensive survey of the variety of tasks English fliers were successfully undertaking fired the young American's imagination. When she flew back to the United States, it was with will aroused to action.

"English women fliers are splendid," she told reporters.

She called them "capable, casually brave, gay, and yet responsible." And she added that it was a great experience to find Government leaders, Army officers, and the Air Force depending on the service women could give. "Here in the United States," she said resentfully, "women fliers are still considered only a colorful and charming addition to the scenery of a flying field."

Jacqueline's longing that women have a real part in aviation during the world crisis was an echo of the emotion felt by Ruth Law and Katherine Stinson during World War I. For a time Miss Cochran may have felt something of the older flyers' sense of frustration. But in the twenty-three years which bridged the two eras the situation of women fliers had basically changed. Now some 2,000 American women held pilots' licenses, and hundreds of them had more hours of flying to their credit than many men. They had demonstrated to the whole world that they had the steadiness, skill, and judgment for important tasks in aviation. Now, therefore, it was possible to get a sympathetic hearing for a plan to organize an auxiliary air corps.

"If this country goes to war," said Jacqueline, "every woman able to ferry planes inside the United States can free a man from this duty for actual combat."

To aeronautical officials, in interviews, and on the air Miss Cochran preached the gospel of service for women fliers. She said they could handle basic teaching of male recruits and serve in many branches of technical work.

When Jacqueline was elected president of the Ninety
Nines, she used her influence directly with the most important group of American women fliers. Meanwhile, she
worked on an immediate plan for action. When Jackie disappeared completely from view, she was more than likely
to be in Canada conferring with Canadian and British
aviation authorities.

Germany's declaration of war on the United States
brought this secret design to a conclusion. Then it was
announced. Under Jacqueline Cochran's leadership a group
of American women pilots, carefully selected by question-
naire and personal interview, was to be sent to England
to serve with the British Transport Auxiliary of the R.A.F.
Captain Norman Edgar, representing that organization in
America, gave out a public statement to express his delighted appreciation of this offer of help.

Jackie set out on a nation-wide tour to push the plan
further, for to her the organization of that unit is only a
beginning. Years ago this dynamic young pilot chafed at
the limitations placed upon women fliers—limitations all
too often complacently accepted by the women themselves.
Training, technique, and experience should be used vitally
and to the full, she believes.

"First you have to have a dream," she said once. "A
dream is a reality. If it doesn't work, take a new angle on
it."

If anyone ever had the right to say this it is Jacqueline
Cochran. Only because of her capacity to "take a new

angle on it" did her dream of being an outstanding flier come true. First was her lack of money and backing. Second was the long series of accidents sufficiently severe to down a flimsier spirit. The final obstacle was set up by marriage to a wealthy man with its inherent temptation to relax and revel in the luxuries of existence. This flier, however, soared lightly over all hurdles.

One photograph of Mrs. Floyd B. Odlum taken in her New York apartment is a real study in character. The background consists of a huge table loaded with trophies. But the young woman who won them is seated on the floor moving airplane models over a huge aeronautical compass protractor set into the parquet. No resting on past honors for her. Action—that's Jackie every time.

Gladys O'Donnell

THAT WINNING WOMAN OF LONG BEACH

☆

I<small>F</small> GLADYS O'DONNELL—she is Mrs. J. Lloyd
O'Donnell to the census taker—chose to wear her medals
on one of those shoulder-to-hip sashes where foreign diplo-
mats display their decorations, she would find the quarters
very cramped. For not only does she enter an aviation race
as casually as most women enter a grocery store, she seldom comes out of one of those races without an award.
According to one authority, she has taken top honors in
twenty-nine competitive events with both men and women.
Her record as a flier of remarkably dependable skill and as
a winner is all the more amazing when you consider that
she began it when she was the mother of two children. Married in 1921 at the age of seventeen to J. Lloyd O'Donnell,
owner of a school of aviation at Long Beach, California,
she hardly thought of learning to fly until seven years later.
In 1928 she took her first lesson in the air.

Of course she had always loved flying as a passenger.
Gradually she came to be deeply interested both in helping

her husband with the vital task of advancing aviation and with flying herself. But in describing her first adventures she always kept on a humorous key. Witness this account published in an aviation journal:

Being quite ordinary people with the usual husband-wife complex, we spent the entire first lesson arguing about which way was up. For the next six weeks I sat with folded hands and watched others fly.

Then it became necessary to hire another pilot to assist with the students and—Eureka, my chance! In a couple of months I was ready to solo. When the great day arrived everybody ran and hid in the hangar while I clattered forth behind the thundering ox. To the astonishment of everybody nothing happened. Nothing unusual ever happened on my flights in those days. . . . That was in 1929.

By a month before the first Women's National Air Derby that same year I had piled up the amazing total of thirty hours. Equipped with this vast experience and a Velie Monocoupe, I decided to fly over the course. It was the best experience I ever had. I taxied as far as Douglas, Arizona. Another eight hours on the trip gave me all of forty-six hours.

And how did she fare—this fledgling—when pitted against the veterans of the air? She came in second after Louise Thaden. With only forty-six hours to her credit, she immediately formed the habit of winning. And though she herself dismissed her exploit as "only a healthy streak of beginner's luck" it placed her firmly in the ranks of aviation.

The derby, as you will remember, was flown from Santa Monica, California, to Cleveland, Ohio. And in Cleveland Gladys decided to tarry for a while. Why not? Weren't the National Air Races now going on in that city? Didn't she have a chance to see the Canadian aviators put on their stunt show and Lindbergh lead the Navy acrobatics and Frank Hawks fly a glider for the crowd?

But she was more than a spectator in Cleveland. Flushed with her success in the derby, she entered several of the events. Two out of three closed events for women she won. One of them—a 60-mile race—she took with a speed of 137.6 miles per hour. Just as easy was her victory in the Cleveland-to-Pittsburgh race.

Meanwhile, what was she saying to the husband who had given her that first argumentative lesson? We can imagine she disguised her excitement with teasing. "How about it, Lloyd? I seem to have found the way up and you can't talk me out of it." Perhaps some such wire traveled to Long Beach.

The following year she kept busy, but it was in no spectacular way. She became one of the expert instructors at her husband's successful school. She worked for her transport license. But her career in the air never interfered with her devotion to her two children, Lorraine May and James L., Junior.

However, the year of 1930 was a different story. Before

it merged into 1931 Gladys O'Donnell must have ex-claimed, "Did anyone say there was a depression? Where's it hiding?" For in that year she was to receive $8,800 in prize money—a total few women aviators have ever been able to boast.

Her lucrative months were launched on June 18, 1930. That day she took off from California for Chicago. Together with her pilot, C. F. Lienesch, manager of the 1930 Derby and head of aviation for the Union Oil Company, she planned to join Amelia Earhart and Louise Thaden, win-ner of last year's derby. The four of them were to scout a route for the women's race. Once this proposed route of 3,600 miles, leading through the United States, Mexico, and Canada, was properly covered she felt free to go back to her winning. On August 1 she came in second at the first Tom Thumb Air Derby—it was flown over a 225-mile course. The winner of this race was Florence Lowe Barnes.

However, this was only an appetizer to Gladys O'Don-nell. The real banquet began when she started out in the Women's Air Derby of that year. Early in the contest it was apparent that she was up to her old tricks. On August 20, for example, it was wired all over the country that she had won both the fourth and fifth laps of the derby. These laps increased her elapsed lead over her nearest com-petitor, Miss Doig of Danbury, Connecticut, to sixteen min-utes. Could she maintain that lead? All interested in avia-tion waited breathlessly.

Three days later the newspapers answered that tense

question. In Wichita, Kansas, Gladys O'Donnell's elapsed time of 9 hours, 55 minutes, and 55 seconds broadened her lead over Miss Doig by slightly more than an hour. Meanwhile, all the fliers had encountered severe winds and rainstorms which forced three of the contestants to land. Meanwhile, too, Miss Doig had developed engine trouble. Finally, in fact, the Connecticut girl was compelled to drop out of the race. And when on August 25 Gladys made the hop from Des Moines to Madison in 1 hour and 41 minutes her victory was a foregone conclusion.

The next day she was to taste the fruits of that victory. The race of 2,245 miles was at an end. And when at 3:18 in the afternoon she crossed the finish line in Chicago cheers burst from the crowd: "Hooray for Mrs. O'Donnell!" And then suddenly through that chorus cut a shrill variation: "Hooray for the red-hot mama." As, beaming and waving, she stepped out of the cockpit she was carried on the shoulders of her admirers to the announcer's stand.

It was indeed Mother's Day in Chicago—the end of the 1930 Derby. For when the runner-up in the race crossed the finish line she turned out to be Mildred Morgan of Beverly Hills. And, though she might not beat Gladys in the air, she won in the nursery. As against the two O'Donnell children there were three little Morgans, two of whom were twins.

By her victory in the derby Gladys won $3,500. It was only a forerunner of other prizes. In Chicago she lingered for the National Air Races. With Gladys, the verb "to linger" is always equivalent to the verb "to enter." And as in her case the verb "to enter" also means "to win" she proceeded to come first in four of the events. These victories included, not only the 800- and 1,000-cubic-inch displacement races, but the free-for-all which yielded both a purse of $2,500 and the trophy donated by Mrs. Robert R. McCormick.

This was the way in which the New York *Times* summed up her prowess in the women's free-for-all:

Winner of this 50-mile event was Mrs. O'Donnell, who gave a splendid exhibition of race flying in her Taper-wing Waco Whirlwind powered. She made the turns like a Doolittle. With the exception of Mrs. O'Donnell, none of the seven contestants was able to make the banks and turns that indicate good aimship.

There were to be echoes of her victory in the 1930 Derby in 1931. On January 27 of that year she was awarded by the National Aeronautical Association the huge and coveted Aerol Trophy. This trophy, which was accompanied by a check for $3,000, was the final seal upon her distinction. She had won the major air event for women in 1930.

In August of 1931 we find her in mixed society. For the first time she entered a race open to both men and women. This derby was flown from the Santa Monica Airport to the Cleveland Air Races, and on the twenty-fourth of the month —have you already guessed who was leading the pack of sixty men and women on its first lap? Why, yes, to be sure, none other than the balky pupil who only three years before had taken her first lesson in the cockpit. In a ship which had the highest handicap—it was at 171 miles per hour— she averaged 162 miles on the 196-mile hop.

Four days later people all over the country were lifting their eyebrows. What in the world was the matter with the men? Why were they lagging behind the women? Although Gladys was leading on that day, August 24, she took up the cudgels for those sluggish males. There was something wrong with the calculations. Handicaps for some planes were not computed accurately.

The derby officials did not agree with her. "Nonsense," barked Mr. J. A. Woodward; "handicapping has nothing to do with the situation. The girls are flying better—that's all." Another woman in the derby agreed·thoroughly with the official. Phoebe Omlie, who was leading the general handicap when the pilots left Amarillo, maintained stoutly that poor navigation explained those backward boys. At all events, the excellent flying Gladys demonstrated in this race was followed by victory in two closed races in the National Air Meet.

The next year, 1932, found her again flying a man-woman race from the West Coast to Cleveland, and again the National Air Races proved an excellent investment for her. She entered the Aerol Trophy Race, which promised a prize of $2,500 to the winner, and wrested victory from a driving storm. Rain deluged the planes of the contestants. The women pilots, seasoned as they were, could not see through the darkness the signaling flag on the pylon. In fact, many of them could not distinguish a pylon from a smokestack. They wandered far afield. One rocketed into the air, another turned up mysteriously from behind a grandstand. Although the race was flagged down after the fourth lap, the contestants failed to realize the order to stop and plodded on through the storm.

Naturally the men pilots were vastly amused by what one of them called the "scramble of the hens to get out of the wet." Yet Gladys O'Donnell could afford to overlook their mirth. In her speedy Howard Special, affectionately known as "Ike," she achieved an average speed of 185.476 miles an hour. That speed meant the $2,500 prize. Her runner-up was May Haizlip, and third in line was Florence Klingensmith, the only one of the contestants who had completed the entire course.

Gladys O'Donnell's succeeding years read like the multiplication tables. There are just more and more victories. For this reason we must telescope these years. In December of 1933 she was one of the two women who finished the Los Angeles Women's and Girls' Air Derby from Los Angeles to San Mateo, and in the derby that followed she came out first in a field of five men in a free-for-all. In 1936, while flying a Menasco-powered Ryan, she placed second in the Amelia Earhart Race at the National Air Races in Los Angeles. In 1937, while flying the same race in the same plane, she bested Betty Browning, the previous year's winner, by a

GLADYS O'DONNELL, winner of many competitive flying events, shown with the Amelia Earhart trophy she won in the National Air Races in 1937.

"Acme"

TEDDY KENYON, a noted test pilot and winner of the $5000 prize as the champion sportswoman, at a national air show in 1933.

speed of 129.653 miles per hour. And these, of course, represent only a few of her conquests.

You would assume that any woman who had such an active career both at home and in the cockpit would find no spare time at her disposal. Yet this energetic flier has always made each day a mere piece of elastic in her hands. She can stretch it to almost any limit. Not only has she been an instructor in the O'Donnell School, but she has been secretary of the O'Donnell Aircraft, Inc., at Long Beach Airport. She is transport pilot 6608 and has, of course, carried a large number of passengers. Also, she is a licensed radio-telephone operator, third class. This alone would distinguish her among other women fliers. It meant special instruction and long practice to attain that rating.

More than this. Here is a woman who has made room in her life for such by-products of an aviator's career as a program on the radio. In 1933 she spoke regularly over the mike in a night broadcast called "Sky Doings." Since heretofore that program had dealt exclusively with the activities of the men pilots, her pioneering counted in woman's march upon the bastions of the air. Of even more spectacular interest is the fact that she was the very first woman ever to fly in the movies. She was one of the pilots in *The White Sister* and she was a marked success. In fact, the director was more pleased with her flying than with that of the men in the cast.

Blessed with a sense of humor which seldom deserts her, Gladys O'Donnell doesn't go about her executive chores with that worried, "How-will-I-ever-get-it-done?" look which disfigures most busy women. For example, in 1935 she was managing director of a Pacific air pageant patterned upon the National Air Races. This pageant included the Women's Championship Air Meet and it offered more prize money than any other similar event that year except the National Air Races. It was a big job. It had a thousand pestiferous details. Yet Gladys O'Donnell took it with the same nonchalance with which in 1929 she entered the Women's National Air Derby to win second place after only forty-six hours in the air.

One of the outstanding fliers of the country—we can leave her only on that same merry note with which she herself has traced her career in the air. In her first lesson she may have been mistaken about "the way up." But, having found the way, she has certainly known how to stay there.

AN AERIAL VIEW OF THE LAST SEVEN YEARS

I<small>N THE</small> seven-year period from 1935 to 1942 American aviation came into its heritage. Air travel changed from an exceptional experience to a commonplace. Sleeper planes skipped across the continent from West to East in fifteen hours. Pan-American Airways began to deliver mail and passengers to Hawaii, the Philippines, and South America. In this period plane design became streamlined, engines more powerful, and safety devices more reliable. Pilots and ground crews increased in both number and skill.

Such a heyday for aviation meant activity for more women fliers. There were four hundred and ten of them in 1936. Although the trails had been blazed by the first generation of women in the air, modern girls were proving worthy of the great tradition. Doubtless each of them has a story worth telling. But here we can only sketch in a few strokes the fascinating picture of what American women were accomplishing in the air.

Let us look first at a small list of special records. It shows the indomitable sportsmanship of the happy amateurs in modern times.

Melba Beard, winner, Amelia Earhart Trophy Race. Cleveland.   1935

Iona Coppegge, light-plane altitude record. Aeronca C-3 plane. Dayton, Ohio. 15,253 feet.   1936

Annette Gipson, first category light-plane altitude record. Lambert Monocoupe, 90 horsepower. Fort Lauderdale, Florida. 12,628 feet. Speed, 123.247 miles an hour.   1936

Helen Frigo, national and international speed record for light planes. 100-kilometer course, Baltimore, Maryland. Speed, 74.193 miles an hour.   1936

Irene Crum, light-plane altitude record. Aeronca C-2 plane. Huntington, West Virginia. 19,426 feet.   1936

Crystal Mowry, seaplane altitude and speed record. Kitty Hawk, 125 horsepower. Altitude, 6,070 feet. Speed, 79.138 miles an hour.   1936

Betty Browning, winner, Amelia Earhart Trophy Handicap Race. Warner Cesna Prize, $675.   1936

Evelyn Hudson, endurance refueling record for solo flight. Oxnard Airport, California. 19 hours, 57 minutes.   1937

Grace Huntington, light-plane, Class-C altitude record. Burbank, California. 16,769.646 feet.   1939

Even these few records culled from the total score made by women shed luster on the national ledger. But records are not the only indication of prowess. The joy and delight women have had in flying reveal just as much about the advancement they have made.

Florence Boswell of Cleveland, a modern pilot who understands new instruments and bad-weather flying, has more hours in the air to her credit than half those who make the headlines. Mrs. Alice Hammond of Detroit likes to pile her children into the back of the plane and go for hops both long and short. Bessie Owens of Santa Barbara went abroad with her plane in 1935, flew all over Europe, and then on to India and China. Eighteen months of flying with never a sign of engine trouble. Another who collected prestige across the seas is Fay Gillis. She flew her own glider in Russia and during a four-year period of residence lectured on American technique at the Aviation Institute of Moscow.

Margaret Cooper, a good flier in her own right, was the first woman in California to own and operate an airport. She is responsible for inspiring interest in aviation among many Hollywood stars, and most of those who have learned to fly give Margaret credit for the impulse. Margo Tanner of Hartsdale, New York, has flown hundreds of hours and won a seaplane record. Mrs. Betty Huyler Gillies, for some time head of the Ninety Nines, is a first-class sportsman pilot. She is one of the very few women who can fly a twin-motor plane; she has also proved herself a good executive for an airplane company. Ruth Chatterton, of stage and screen fame, has been flying just for fun ever since she had the money to buy a plane. The Ruth Chatterton Sportsman

Pilot Air Derby she established paid $2,000 in prizes when last held.

These are mere samples of the 1,540 licensed women pilots actively flying in the United States. Some of them as yet have only solo licenses for eight hours in the air. Some have the private license requiring thirty-five hours' solo work. A small number possess the commercial license based on two hundred solo hours in the air. Every one of these pilots helps maintain a high level of general interest in aviation and adds to the rapidly increasing experience upon which youthful fliers of the future can draw.

When it comes to gliding, however, our American women will have to take to pioneering once more. For this type of conquest they have not shown the same zest as do their European sisters. As we have said in another place, Anne Lindbergh was the first woman to get a first-class glider pilot's license requiring six minutes in the air. That was in 1930. The next year two other women received gliding licenses, and a flurry of interest resulted in the forming of two gliding clubs, one in California and the other in Buffalo, New York. The first president of the California club was Maxine Dunlap Bennett, who took her flying license in 1928 and has established a world speed record for light planes.

The great Eastern gliding center of the country is at Elmira, New York. High hills and deep valleys in the beautiful region offer excellent conditions for both take-offs and soaring. Gradually Harris Hill, with its level top and steep incline and its nearness to the city, became the official gliding field. Hangars and an administration building offer good accommodations. It is there that records are established, and until World War II broke out Harris Hill was the scene of a distinguished international gathering.

With a few exceptions women joined this group only to applaud the skill of men. In 1931 Dorothy Holderman remained aloft for 45 minutes. In 1938 Helen Montgomery of Detroit made a national endurance record for women of 7 hours, 28 minutes in her glider. Compare this, however, to the 24 hours, 14 minutes' record of a certain Polish woman, and you see why these feminine achievements drew little attention. Here is one more indication of a fact which is now being forcibly presented to American military and naval commanders. Never has the glider, either as a means of training youth or as a useful adjunct to war maneuvers, been recognized as important in the aeronautical circles of this nation.

It is when economic necessity combines with interest in flying that women show their greatest adaptability. In 1929, just thirteen years ago, there were ninety-nine women fliers in the land eligible for the Ninety Nines. Today, in addition to all the women who fly for sport and those who work for the Government, one hundred and twenty-three women hold commercial positions in aviation. From Fay Gillis, an aviation editor, and Barbara Archer, traffic manager of Northwest Airlines, to Marty Bowman, veteran contestant in air races, who ferries ships from factory to destination points, these professional aviators have shown a highly diversified activity. Each personal history adds a fascinating bit to feminine annals.

Take the story of Laurette Schimmoler of Ohio. Hardly had she secured her pilot's license in the late 1920s when she started a very unusual enterprise. She owned and operated a municipal airport at Bucyrus, Ohio. In connection with it she founded an institute of aviation. Neither project was on a large scale, but Laurette might have been content to develop them both, had it not been for an experience which gave her life quite a new direction.

One summer a terrific tornado ripped through the unsuspecting state of Ohio. Shortly afterward Miss Schimmoler was flying over Lorain, a town which had lain in the direct path of the hurricane. Her horrified eyes beheld all the signs of destruction and her imagination was jolted into action. She was immediately fired with the idea that aviation should be directly useful to victims of catastrophe. "What we need is a Red Cross of the air!" she said.

Recognizing this as a real inspiration, she set to work to make it practical. To begin with, she realized that she herself must have more knowledge of aviation. First, she managed to secure a job as ground employee with the United States Air Mail Service in Burbank, California. To this training she added two months' experience with the U.S. Airport Weather Bureau and topped it off with fifteen months' work at the Lockheed Aircraft factory. During this period of education she developed contacts with nurses and the medical profession. Among these groups she found many individuals ready to support her plan.

At last, in 1936, Miss Schimmoler founded the Aerial Nurse Corps of America. Its headquarters, where she maintains a national center of contact with member nurses, is at Burbank, California.

The organization offers the usual honorary and associate memberships to all interested supporters. But active membership has specific requirements. To qualify, a trained nurse must be between the ages of twenty-one and thirty-five and must be able to pass severe physical and professional tests for air duty. For its operation, the Aerial Nurse Corps slices the nation into three main sections called "Wings." Each Wing has various "Divisions," and in some major city is located the Division Headquarters. Any calls for aerial nurses can, therefore, be filled. Probably the airports have chiefly benefited from the service. But medical aid has also been flown to isolated places, some accidents have been covered, and nurses have accompanied patients who were flown to clinic or specialist. The National Aeronautics Association has recognized the Aerial Nurse Corps as an official nursing organization.

It is inspiring to find an aviator like Laurette Schimmoler who can succeed in realizing a purposeful plan. Most fliers have grasped opportunity rather than created it. Sometimes, however, stark necessity shapes an aerial career. That was the case with Mabel K. Wilson. She suddenly had to face a desperate situation, and the story of her doing so is one no novelist could better.

Mae, as she was known to many, was a young woman completely absorbed in the career of her husband, Roy Wilson. He was manager of his own private flying field in the northwest outskirts of Chicago. Here he gave flying lessons and built up an excellent aviation unit. He and Mae lived at the airport. Sometimes they had plenty of money and sometimes very little. But they loved the ups and downs of existence, their hops to Florida, the nice boys who took lessons, and the racily good-natured ground crew.

Then one terrible day Roy Wilson was killed in a crash. With happiness and source of income gone, Mae Wilson confronted a set of circumstances tragic enough to daunt a lesser spirit.

"Sell the flying field and the two planes, Mae!" That was the gist of the advice offered her by family and friends.

But Mae refused to listen. "I'm going on with Roy's work," she said stubbornly. "I intend to operate this field myself."

In vain she was reminded that she couldn't fly and had little capital for operating expenses. With impassioned resolution she announced that she was going to learn to fly, that she meant to engage pilots as instructors and stay just where she was.

For a long time, however, she met nothing but discouragements. Men didn't like to work for a woman, and pilots were not to be hired. One student after the other dropped away. She was almost at the end of hope when a Canadian flier named Marshall O'Neil came to the airport. Calmly he took hold of the situation. He taught the remaining students and gave Mae Wilson herself a rigid course of instruction. She was not one of those natural fliers, but with hard, intelligent work she qualified to pass her various tests. But the school was still so small that expenses could barely be met. To the young males of the Midwest a flying field operated by a woman lacked virility.

Apparently it was a narrow escape from death which banished prejudice against courageous Mrs. Wilson. She was flying an old Travelair plane with a student who wanted to learn stunts. In the midst of the maneuvers the young man jammed the controls and caught his foot against the stick in the floor of the cockpit. He couldn't wrench it free. Mae pulled and pushed, but failed to budge either foot or lever. The boy was frantic, but Mae remained calm. They were still high enough to have a small margin of time.

"I'll have to rip the floor boards," she said.

It was a task for a strong man. But somehow Mae managed to smash the wood and loosen the imprisoned foot. In a moment she had leveled off the ship and made a perfect landing. As she saw the amazed admiration and gratitude in the boy's eyes, the pilot realized she had won more than the fight for both their lives. She knew that at last she had qualified as a flier.

From that moment on her reputation was made. The story made the rounds, and both students and private passengers began to flock to the Wilson Flying Field. In 1939 Marshall O'Neil was obliged to leave his work because of ill-health. But in five years of piloting for Mrs. Wilson he had seen her through to the end of the hard struggle and knew she could carry on. Such resolution as hers helps establish confidence in all the oncoming fliers who know Mae's wonderful history.

Many times in this book we have described the demonstration of planes by various women fliers. But the woman engaged in exact commercial testing of new models or new installments is a rare bird, indeed. Up to the present only two women have secured this type of commercial work.

One of these exceptions is Nancy Harkness Love. She is the wife of Robert Love, who was connected with the Waco Airplane Company. Theirs was one of those winged romances which began when Nancy undertook to demonstrate the Waco planes. Member of the Ninety Nines and jaunty participant in many racing meets, Nancy was soon a marked figure in aviation. With Louise Thaden and Helen McCloskey, she was one of the trio first invited to work as air-marking pilots for the Bureau of Air Commerce. That was in 1934. Nancy covered every single city in New York State to make surveys of the best air-marking sites, and for each project she set up the plan.

In 1937 the girl left the bureau to serve as salesman-demonstrator for the Gwinn Aircar Company of Buffalo,

New York. At the Cleveland Air Races that year the designer's new ship with its many novel features was shown by Nancy Love and by the great racing pilot, Frank Hawks. Hawks, who served as commercial test pilot for the Gwinn Company, was called out West, and Nancy offered to put the plane through its tests until he returned.

"I've been flying the Hammond plane for the Bureau of Air Commerce," she remarked to Joe Gwinn, "and that isn't a conventional type either. I think I can put the Gwinn through its paces."

It was a courageous proposal. The Gwinn plane was designed with only two controls and had no rudder. It was steered just like an automobile, and its elevator control was limited in order to prevent the ship from being stalled or spun or nosed down into a steep dive. The landing gear was of tricycle type and the ship was supposed to land with the control wheel in any position.

"We're going to test the landing gear," Gwinn said to his new test pilot. "We want to see how much punishment it can take without smashing the undercarriage."

Nancy felt a little faint at this prospect. But when the morning of the test came, she was there looking adequate and unafraid. To get the quietest air, the test was made at dawn. Staring into the cockpit, the girl made out in the half light a weird collection of recording instruments and an accelerometer installed on the floor at the center of gravity. The luggage compartment was packed with bags of shot to bring up the ship to full gross weight. Joe Gwinn strapped Nancy firmly into the pilot's seat. Then, armed with a check list, a stop watch, and a slide rule, he seated himself beside her to take the readings.

Test number one was a glide from 1,000 feet. Never had Mrs. Love made a descent under such conditions. The flaps were to be down, the motor throttled back for a landing, and the wheel full forward. In the usual airplane this would be an effective means of committing suicide. At 1,000 feet, with the throttle back, Nancy had to use all her will power to push the wheel forward. It went against everything she had ever learned or practiced in seven years of flying. The plane shot down at 980 feet a minute. Afterward Nancy vividly recorded her sensations:

The last two hundred feet offered the worst experience I've ever had in my life. The ground came up at us with unbelievable speed. My insides felt sort of shriveled up and I know I was holding my breath as we dropped like a plummet onto the runway. The impact was tremendous, as I could tell from the almost explosive sound of the oleo shock absorbers taking up. But, though I had expected to be pushed bodily through the seat, I only felt a very small jar.

I don't remember taxiing in. I must have been almost unconscious from the shock of fear that had preceded the impact. Then Joe said in what I thought was a weak tone, "That was fine." I didn't answer him. I couldn't.

The pilot hoped such an effort was enough for one day. But, oh no! Up she had to go to repeat the test. This time

LAURA INGALLS on her return from her history-making aerial circuit of South America.

NANCY HARKNESS LOVE, one of the best-known test pilots, with the Gwinn "Aircar," one of the many types of planes she has demonstrated.

she could not make that effort of will to push the control wheel forward. At two feet off the ground she pulled the wheel back for a normal landing.

Gwinn looked at her reproachfully. "You've got to get over that, Nancy," he said. "The ship is designed to make the landing we're testing her for and you've got to let her do it."

Let her she did. Over and over again that day the same nightmare was repeated. Between trips, while mechanics made inspections and filled the tanks with gas, while Joe Gwinn compiled data, Nancy lay on the grass and groaned inwardly. At the end of the day, completely exhausted, she staggered home to bed. Now she knew at first hand what paces every new plane model goes through in order to prove itself safe for public use. Never again, she told herself, would she take for granted that unknown hero, the test pilot, risking his life in the course of the day's work.

Nancy's second day went a little better. She couldn't help a sickening feeling of falling through space, but she stood it with more faith in the little ship's ability. She wound up the afternoon by taking the secretary of the company for a quiet spin over Buffalo and slid gracefully down for a normal landing. Instead of the gentle contact expected, however, there was a loud bang as they touched earth. Something flew up on the left side. The plane tipped with left wing dragging on the runway and the right side of the plane high in air. After spinning in a circle with a grinding, tearing sound, the plane came to a stop.

In the awful silence the girl thought, "Poor Joe Gwinn! What have I done to his plane?"

There he was running across the field with his engineer and mechanic to survey the damaged plane. Too curious to be disheartened, Gwinn began to search for the cause of the mishap. It was soon found. A broken casting in the landing gear had caused the accident. Nancy was told that if she hadn't made a soft landing that time the ship might have been completely wrecked. She almost cried with relief.

For two weeks the determined girl continued to fulfill the difficult assignment. Worn out, but uncomplaining, she piloted the Gwinn plane day after day without accident or failure. Although at the fortnight's end she concluded that expert testing of so violent a kind was hard on the feminine nervous system, she always felt she wouldn't have missed the experience for worlds.

Nancy remained with the Gwinn Company for almost a year. The routine testing, flying, and demonstrating she was given offered no strain to a pilot of her skill. This, indeed, is the work she likes best, and she has followed it with a number of airplane companies. Into the pilot's seat of a new plane she slips with more nonchalance than most women feel when asked to drive a slightly unfamiliar motorcar.

Quite a different story of testing is presented by Mrs. Theodore Kenyon. Since she is generally known as Teddy Kenyon, her husband has to shorten his name to Ted. He was first a barnstorming flier, then a transport pilot, and finally a technician for the Sperry Instrument Company. Not only did he teach his wife to fly, but continued to coach her long after she was skillful enough to enter contests.

Probably it is the wonderful sympathy between Ted and Teddy which has guided the latter's career to its present status. Nineteen hundred and thirty saw its beginning. Mrs. Kenyon entered an unusual contest for a $300 cash prize offered women fliers by the American Legion. Each competitor had to execute exact maneuvers and landings of sufficient difficulty to test her skill. In addition, she was to be rated for appearance and manner. Conscientious practice under her husband's critical eye assured Teddy that she could meet the test in maneuvers. The rest was her own affair. Born with charm of manner and appearance, she needed no practice in smiling poise. All the clever young woman had to do was to don a well-tailored green costume to match her green plane. Inevitably she returned the winner.

Ted, however, was not impressed. "That's fine, honey," he said; "but now I'm sure you want to fly in earnest."

For three years she went to air meets up and down the Eastern coast. Boston, New York, and Florida contests were sure to find Teddy Kenyon on the list of entrants. Whenever Ted saw his wife fly, he would go over every move with her afterward, draw diagrams, and explain how she could improve her technique. Because that was exactly what she wanted, she never stopped learning.

Typical of this attitude was her approach to a noteworthy

event in 1933. A charity air pageant of vast size was to be held on Long Island. Sportsmen, Army and Navy aviators were to put on a thrilling show. As a finale a champion sportsman and a champion sportswoman were to be selected by the judges. Each winner received a $5,000 prize. As prelude to the event a cross-country treasure hunt was held. Clues were strewn from Albany to St. Louis, and the fliers anticipated a lark not only in the search for clues, but in the sociable sessions after the day's hunt.

It was a tempting program. But when Teddy discussed the affair with Ted, she remarked with great decision, "I'm not going to join the treasure hunt. That's only a side show. I'm going to practice for the big feature. It's no cinch."

"Good girl!" approved the perfectionist husband. "What are you going to use for a plane?"

Ah, but she had thought that out, too. A man she knew in Boston had a good reliable Waco plane with a 120-horse-power Warner engine. She struck a bargain with him. He would lend her the plane, and if she won the prize, she would give him $1,000.

For days on end, while many of her friends winged their merry way on the treasure hunt, Teddy Kenyon worked harder than ever in her life. Time after time the little ship would circle in the air, point at the field, and glide down to land on a predetermined spot. Time after time the Waco dived, pulled up to perform a loop, and then completed one Immelmann after the other. Following the requirements of the coming event, Teddy would stall, then slip into a tight spin—round and round—until time to pull out and level off. When the day of the meet arrived, Teddy flew her plane to Roosevelt Field with a new confidence borne of mastery.

The charity pageant had drawn all the notables in aviation. Records were broken. Marvelous stunts were performed. Army and Navy maneuvers were of a kind to thrill the most sophisticated. But private pilots found the real excitement far from the grandstand. Thirty-nine men and women were competing for the sportsman's championship. The tests, although neither dangerous nor extraordinary, were of a kind to demand such skill and precision that only those who had practiced every move stood a chance of winning. One by one the contestants dropped out, until only four men and four women were left. Again these eight fliers went through the program—three landings to a mark, two loops, two precision spins, two snap rolls, and two Immelmanns. At last four planes rose into the air. After that trial it was all over, and the woman champion was Teddy Kenyon.

One of the many stories about the winner described her use of the $5,000 check. She cashed it in the morning—so ran the tale—and by night she had paid off her debt, bought an automobile, an airplane, three suits for her husband, and a few odds and ends for herself. Rich, but penniless, she went home happy.

Fame won by supreme skill is the best possible professional asset. Because his wife's ability had been so often judged excellent by experts, Ted Kenyon had no hesitation in engaging her to assist him. As copilot and pilot she helped him in his experiments for the Sperry Instrument Company. First she demonstrated the Deviometer, a new blind flying device. Then she flew directed courses for the Sperry anti-aircraft locators. In the summer of 1937 Ted began working on the development of an automatic pilot for the Sperry Gyroscope Company and, believe it or not, he kept his same assistant. Next year Teddy herself flew to the Wright Field at Dayton, Ohio, to demonstrate the automatic pilot to Army officials.

Anyone watching this slim young thing playing tennis or dancing or doing a jackknife dive would naturally think she was only a sportswoman. But beneath the gay disguise lives a student of navigation and a good mechanic. The Ninety Nines likes to boast of Teddy's knowledge of aviation science. But she hands the credit to her husband. Never was a more congenial pair than these Kenyons. Whether at work together on intricate calculations in the shop, or skimming the skies just for a look at the sunset, they share a basic joy. Both of them love to do a thing well and then learn to do it better.

Another type of work still exceptional among fliers is aerial photography. Here again only two women have taken it up. One of them is Bernice Blake Perry of Wilton, New Hampshire, and the other is Jean Adams Cook, one of the authors of this book.

Camera work of this sort requires adequate flying skill. It

must not be confused with shots made on the ground. General photographic publicity for aviation employs the regular technique of picture taking, and many women are engaged in it. Magazines and newspapers when we are not at war are always eager to use pictures of new planes, of airports and factories, of famous fliers and passengers. These are all ground shots.

A famous promoter who directs the taking of such pictures for her firm is Miss Patricia O'Malley. She heads the Traffic Promotional Department of a nationally known air line with an office in Washington, D.C. Covering in normal times as much territory as if she rode on a magic carpet, she herself has been widely publicized.

Geraldine Carpenter, on the other hand, is herself a photographer. Connected with an Eastern air line, she has headquarters in Montreal, Canada. Her duties consist of traveling over the company's line to make graphic records of its operations for advertising. One day tiny Geraldine perches with her camera high on top of a shining Douglas fuselage to photograph passengers boarding the ship. Another time she might lie flat on the ground to get a shot through the blades of a giant propeller outlined against the sky.

All such work is interesting. If the women who do it happen to be fliers, they are especially qualified to select the best shots for publicity. Aerial photography, however, can be taken up only by women well versed in flying technique. Moreover, in addition to possessing all the skill required for ground photography, these specialists have to be ready for acrobatics.

The photographer is seated in the back of the small cabin plane usually flown in camera work. Her perch, with feet hanging out into space, is where the side cabin door ought to be. The door is removed to give her free scope. A venturi tube, which will suck out the air in the camera and leave the film flat, is tied to one of the struts. The big fifty-pound camera is tied to the floor and the operator, also tied in place, points its nose out and down toward the ground. A picture made in this way is called "an oblique." "Verticals" are made through a hole in the bottom of the plane at a height of ten or twenty thousand feet. Such photographs, taken in a precisely timed straight line, are made up into sectional scaled maps. They require a specially equipped plane, and both photographer and pilot must be provided with oxygen tanks.

Far more usual is the oblique aerial picture. Clad in flying suit and boots, the photographer waits until the pilot has climbed to 1,500 or 2,000 feet before starting to work. The difficult part of the job is to look through the sighter of the camera and at the same time direct the pilot where to go and at what angle to put the ship. No one could direct such operations who had not herself piloted a plane. Moreover, a photographer must be prepared in emergencies to take over the controls. Successful pictures demand that the plane be held very steady and that the struts, wing, and tail surfaces do not appear on the film. Some danger is connected with this undertaking. For most of the shots wanted demand slow glides over high city buildings and congested districts, over factory sites and crowded streets. A pilot enthusiastically avoids all these areas in normal flying.

Far more enjoyable are assignments to take country estates, country schools, yachting and harbor scenes. Whenever possible, aerial photographs are shot at high noon, when light is brightest and shadows fewest.

What moments of beauty reward the flying photographer is revealed by this description of an actual experience:

One warm summer afternoon, when for once I was acting as pilot, a thunderstorm had just passed by. Shining through great puffs of clouds, the sun lighted their depths from within. The sky was blue as Bermuda water and iridescent clouds floated against it.

All at once, as if we had suddenly touched the Promised Land, a rainbow arched before us. The photographer signaled to me to bank to the left. As I did so, I saw ahead of me a picture I shall never forget. Splendor was all about us, and below in the rainbow itself fell the shadow of our plane.

The phantom ship beneath glided through an arc of colors surrounded by magnificent mountains of shining cloud. Above that ship of shadow were we, in a real plane, alive. It was incredible. And all about us was glory. It was like the glory of God.

Just as a sense of beauty is needed by the aerial photographer, so patience is required by the teacher. The flying instructor has to make Job at his tent door or a cat at a mousehole look vacillating. Explain and drill and then explain again! Unless a woman enjoys answering questions and exploring talent, she'd better not enter this occupation. But in general women are patient creatures. From the great Marjorie Stinson, our first "schoolmarm of the air," to the young fliers of today, who put army recruits through their first lessons, American women fliers have taught well.

There are numbers of women now so occupied. Most of them do so in a private capacity. But at present three outstanding musketeers have been booted and capped for Government service. Tentatively Uncle Sam is trying out the novelty of employing girls to teach men the basic skills of air service. These pioneers are Gertrude Meserve of Boston, Barbara Kibbee of New York, and Elsa Gardner of New Orleans.

To qualify, each of the trio had to have a commercial license and an instructor's rating. The schools where they teach must be officially approved. All over the country such centers have been selected for training civilian youths as pilots. Much pressure has been put on the Government by noted women fliers to make more extensive use of the hundreds of licensed women pilots for teaching. If war brings this about, a lad may soon find no more novelty in being taught the rudiments of aviation by a woman than the traditional three *R*s.

CONCLUSION AND PROMISE

☆

Thirty years of experience stretch behind the woman flier of today. What it reveals is the valiant conquest of handicaps. Flying has not been made easy for women. For the tremendous cost of airplanes, of instruction, and contest fees has heaved up a mountain over which few have soared. The girl who rose to a place of mark in the air without money or special backing is rare as genius.

A review of the histories presented here makes clear this truth. Most of the women who were able to get the essential tools for success in aviation had married pilots or promoters or men of wealth. Remember that only a handful of women have found it possible to make a living by flying. This is because men have not welcomed their competition. Moreover, for all its eagerness to play up the drama of women in the air, the world has an age-old distrust of feminine stability and endurance. Nor is this world composed of men.

How many women passengers would travel on a big air liner staffed with pilots and navigators belonging to their own sex? Can't you hear the Park Avenue duchess who dedicates her trust to men only crying, "Do you suppose I would set foot in that plane? Why, it has a woman pilot!"

Proof of the very qualities which the duchess demands for her safety has been set forth in these pages. It is a happy fact that every instance of clear-headed courage and control on the part of women pilots chisels away a bit of prejudice against their true worth. Yet to establish themselves as economic assets to aviation is slow work, and women must be patient. For some time to come they will have to be content to manage airports, serve as traffic officers in air lines, as editors of aviation magazines, as receptionists, clerical workers, stewardesses, nurses.

But, you may ask, what about all the girls who have demonstrated planes? There have actually been few so occupied. A demonstrator is a sales person. The most skillful pilot is not retained by an airplane company unless she sells planes to customers. Just now, in wartime, of course, when sports flying is banned, the sale of private planes has all the radiant possibilities open to automobile salesmen.

Indeed, military necessity leaves exactly two good openings for the woman who wants to earn her living as a flier. One is ferrying planes from factory to training station or shipping point. The other is teaching. Should our national resources in women aviators be used to the full for these two services, the future may present far more varied opportunities. This is a crucial period of test.

The great news is that women are ready to meet it. No

longer do they stand helplessly by while mechanics warm up their planes. The girls know how to tune up their engines, service their planes, and make many minor repairs. They understand navigation and the use of modern instruments. If they are given the chance to demonstrate their skill on a grand scale, the postwar years ought to offer plain sailing.

To say that, however, is not to suggest an absence of trails to blaze. Quite the contrary, there will be more than ever. For when expertness is taken for granted, initiative comes into its own. Thirty years ago Harriet Quimby's flight across the foggy English Channel was miracle enough. Tomorrow, when ships almost fly themselves, the distinguished woman pilot will be she who devises means of making aviation better serve mankind.